**New Insights into Fundamental Physiology
and Peri-natal Adaptation of Domestic Fowl**

New Insights into Fundamental Physiology and Peri-natal Adaptation of Domestic Fowl

Edited by

S Yahav and
B Tzschentke

NOTTINGHAM
University Press

First published by Nottingham University Press
This reissued original edition published 2023 by 5m Books Ltd www.5mbooks.com

British Library Cataloguing in Publication Data
New Insights into Fundamental Physiology and Peri-natal Adaptation of Domestic Fowl
I. Yahav, S. II. Tzschentke, B.

ISBN 9781789182941

Disclaimer

Every reasonable effort has been made to ensure that the material in
this book is true, correct, complete and appropriate at the time of
writing. Nevertheless the publishers and the author do not accept
responsibility for any omission or error, or for any injury, damage, loss
or financial consequences arising from the use of the book. Views
expressed in the articles are those of the author and not of the Editor
or Publisher.

Typeset by Nottingham University Press, Nottingham
Printed and bound by PODWW, UK

Preface

The workshop on *'Fundamental Physiology and Perinatal Development in Poultry'* was organized by the World Poultry Science Association – Federation of the European Branches – Working Group of Physiology, and the Working Group of Perinatal Adaptation at the Institute of Biology, Humboldt-University of Berlin.

The aim of the symposium, which was the second one organized by the two working groups, was to stimulate exchange of ideas and information on topics related to early development of body functions in poultry and fundamental physiology in general. A further aim was to encourage students to participate and present their studies. Indeed 22 students attended, presenting papers or posters.

Eighty participants from 17 countries attended the symposium. Thirty six oral presentations and 22 posters were presented on the following topics:

- Metabolism during embryogenesis and post-hatch
- The cardiovascular system – embryogenesis and post-hatch
- Different aspects of development
- Fundamental physiology
- Epigenetic adaptation
- Muscle development
- Cellular and molecular aspects of early development and epigenetic adaptation

At the end of the symposium, a round table discussion on *"Alterations During Chick Embryogenesis – Does it Meet the Poultry Industry Needs?"* was held.

The book includes 19 representative manuscripts from different topics presented in the symposium partially summarizing the new insights into fundamental physiology and perinatal adaptation of domestic fowl.

The next symposium will be held in Berlin, Germany from 5th to 7th of October 2007.

PD Dr. Barbara Tzschentke
Prof. Shlomo Yahav Ph.D.

TABLE OF CONTENT

Cellular aspects of early development of the thermoregulatory system

Different aspects of development during embryogenesis

Different issues in fundamental physiology

Early development of cholinergic heart rate control in embryos of broiler and White Leghorn chickens

H. YONETA, S. FUKUOKA, R. AKIYAMA AND H. TAZAWA*

Department of Electrical and Electronic Engineering, Muroran Institute of Technology, Muroran, 050-8585, Japan

In order to examine whether the cholinergic chronotropic control of heart rate (HR) in chicken embryos is different between strains with regard to its development, instantaneous heart rate (IHR) of embryos in two strains, broiler and White Leghorn, was measured simultaneously at the same ages and in the same measuring chamber. Fertile eggs of both strains, which weighed statistically identical, were incubated simultaneously in the same incubator, and 10 embryos of each strain were determined for IHR during a 1-hour period by electrocardiogram on Days 10, 11, 12, 13 and 14 of incubation. On Day 10, IHR baseline was flat for 1-hour period in all 20 embryos measured. Small decelerations of IHR began to appear in some embryos of both strains on Day 11, and all 20 embryos examined on Day 12 had transient HR decelerations with subsequent increases in magnitude and frequency on the following days. As a result, HR decelerations appeared at almost the same developmental stages in embryos of both strains. Power spectral analysis of IHR also showed no particular difference of spectral power between strains, suggesting that development of HR decelerations is identical in broiler and White Leghorn chickens. However, mode HR (MHR) during the 1-hour period in broiler chicken embryos was significantly higher than that in White Leghorn embryos on any days tested (Days 10-14). Shell gas conductance determined in a separate experiment was also different between both strains. However, it is unlikely that MHR difference is attributed to the difference of shell conductance.

Keywords: cholinergic chronotropic control, strain-difference; broiler; White Leghorn; embryonic heart rate; mode heart rate; heart rate deceleration; shell conductance

Introduction

In human foetuses, an abrupt drop in heart rate (HR) appeared in the 20- to 22-week gestational period (Sorokin et al., 1982). The drop in foetal HR occurred within one or two observed heartbeats and recovered to its baseline value within 10 sec. The abrupt decrease of at least 10 beats per min (bpm) below baseline was categorized as HR deceleration (Sorokin et al., 1982). In chicken embryos (broiler strain), beat-to-beat heart rate, that is, instantaneous heart rate (IHR), begins to fluctuate with an appearance of transient deceleration

*Corresponding author: TAZAWA, Hiroshi
E-mail: tazawa@mmm.muroran-it.ac.jp

during the late second week of incubation and the HR deceleration increases in magnitude and appearing frequency with embryonic development. By atropine administration through the allantoic vein, the transient HR decelerations have been known to be mediated by the vagus nerve function and the vagus is already tonic during the late second week of incubation in the broiler chickens (Höchel *et al.*, 1998; Chiba *et al.*, 2004). Controversially, it has been reported that White Leghorn chickens are lacking in the cholinergic chronotropic control during the entire period of embryonic development (Crossley and Altimiras, 2000) and bantam chicken embryos have obvious cholinergic tone (Crossley *et al.*, 2003). Another study with Hisex white eggs reported that parasympathetic activity to modulate embryonic HR already reached a constant level on Day 19 of incubation (Aubert *et al.*, 2004). It is possible that the inconsistency between the reports on the cholinergic chronotropic control of HR is due to the different strains used in the different experiments. Accordingly, in order to compare development of transient HR decelerations between different strains, we designed an experiment to measure IHR of embryos in two strains; broiler and White Leghorn chickens, simultaneously at the same age (Days 10, 11, 12, 13 and 14 of incubation) and in the same measuring chamber. Additionally, according to a suggestion given at the workshop (The 2nd Combined Workshop Fundamental Physiology of the European Working Group of Physiology and Perinatal Development in Poultry) that eggshell gas conductance might be different between both strains and affect embryonic HR, we approximated afresh eggshell gas conductance taking advantage of the data of egg mass loss during incubation which had been determined preliminarily in a separate experiment.

Materials and methods

Fertile eggs of broiler and White Leghorn chickens were obtained from a local hatchery and Hokkaido Animal Research Center, respectively. Eggs were weighed to 0.1 g and numbered on the eggshell for identification. Eggs of both strains were simultaneously incubated in a forced draft incubator that was made specifically for laboratory experiment (inside dimensions; W60cm, H50cm and D50cm) at a temperature of $38\pm0.3°C$ and relative humidity of about $55\pm3\%$ until the experiment began. Eggs of broiler chickens were placed one side and an equal number of eggs of White Leghorn chickens were placed another side of the incubator and turned every 3 hours. On the day of the experiment, three needle electrodes for electrocardiogram (ECG) which were made from 24-gauge hypodermic needles 2.5-cm long were inserted into the egg and fixed on the eggshell. Two eggs of the broiler and another two eggs of the White Leghorn chickens were placed in a measuring chamber warmed at 38°C and ECG was measured simultaneously in four eggs. The amplified and band-passed ECG signals were sampled at a frequency of 12,000 Hz by a 16-bit analog to digital converter (sound card) (Khandoker *et al.*, 2003, 2004). IHR in beats per min (bpm) was calculated with the aid of a computer from the time interval of QRS waves as mentioned elsewhere (Moriya *et al.*, 1999, 2000). The sequence of individual values of IHR was displayed by the sequence of individual points on a computer monitor. After at least a 1-hour period for temperature equilibration and HR baseline stabilization, IHR data were stored on a computer file. The simultaneous measurement of embryonic IHR in the broiler and White Leghorn chickens was made on Days 10, 11, 12, 13 and 14 of incubation.

In preliminary experiments to practise measurement of ECG, we had measured the egg mass of the broiler and White Leghorn chickens, not only prior to incubation, but also before the measurement of ECG on Days 10, 11, 12, 13 and 14 of incubation. Using these data, we approximated water vapor conductance of the eggshell (GH_2O in mg•day^{-1}•torr^{-1}) in individual eggs. The loss of egg mass (in mg) was divided by incubation days and water vapor pressure difference between the egg and the air in the incubator; i.e., 22.2 torr, corresponding to GH_2O. GH_2O in mg•day^{-1}•torr^{-1} was converted to oxygen conductance (GO_2 in ml•day^{-1}•torr^{-1}) by multiplying by a factor of 1.06, because 1 mol water vapor (molecular weight, 18.02 g) is equivalent to a molar volume of 22.414 liters and GO_2 is related to GH_2O by the diffusion coefficient ratio ($DO_2/DH_2O=0.23/0.27$) (Hoyt *et al.*, 1979; Tullett, 1984).

IHR deceleration was defined as a transient decrease of more than 10 bpm during a short period of time. The period was decided according to the magnitude of the decrease in order to distinguish from artifacts; that is, 5-sec period for a decrease less than 20 bpm and 10-sec period for a decrease more than 20 bpm.

A one-hour segment of IHR was taken from the continuous recording in individual embryos. The value of HR that occurred most frequently during the 1-hour segment of IHR, corresponding to mode heart rate (MHR), was calculated instead of the mean values as mean values might be affected by outliers; i.e., HR decelerations. For the same 1-hour segment, power spectrum density (PSD) was calculated by fast Fourier transform. PSD was divided into two frequency regions; that is, the region lower than 0.1 Hz (low frequency region, LFR) and the region higher than 0.1 Hz (high frequency region, HFR). Spectral power (SP) of the frequency was calculated for LFR and HFR in individual embryos. All the values are shown by means with standard error (SE).

The differences of mean values between two strains and between incubation days were examined by two-way factorial ANOVA with repeated measures. When the difference of means between incubation days was significant, Tukey post hoc test examined on which incubation days the means were significantly different. The significant level was P<0.05.

Results

EGG MASS

Ten eggs of broiler and White Leghorn chickens, respectively, were measured for ECG on Days 10, 11, 12, 13 and 14 of incubation. Table 1 summarizes fresh egg mass of the broiler and White Leghorn chickens used on individual days of the experiment. The differences between broiler and White Leghorn chickens (50 eggs each) and between incubation days (20 eggs each) were analyzed by two-way ANOVA, revealing that fresh egg mass between the broiler and White Leghorn chickens was not different (P=0.472) and egg mass on individual incubation days was also different (P=0.910). Table 2 presents mass of the eggs used in the preliminary experiment which provided eggshell conductance. The mean fresh egg mass of both strains (91 eggs each) was not different (P=0.546) and egg mass on individual incubation days was also not different (P=0.890).

MEASUREMENTS OF INSTANTANEOUS HEART RATE

Examples of 1-hour recording of IHR on individual incubation days and PSD are presented in Figure 1 for broiler chickens. In a 10-day-old embryo, no particular HR decelerations of more than 10 bpm from the baseline appeared. PSD of the frequency composing the 1-hour recording of IHR was flat in HFR. In an 11-day-old embryo, three small HR decelerations exceeding 10 bpm appeared during the 1-hour period and PSD slightly increased. Another four embryos

Table 1. Fresh egg mass (g) of broiler and White Leghorn chickens.

Days	10	11	12	13	14	Total
Broiler	65.9	65.1	65.8	64.0	65.2	65.2
	±1.3	±0.8	±0.9	±1.2	±1.7	±0.5
White	65.7	64.3	66.4	66.7	65.4	65.6
Leghorn	±0.8	±0.8	±1.1	±1.4	±1.3	±0.5

Mean±SE, N=10

Table 2. Fresh egg mass (g) in preliminary experiments. Mean±SE.

Days	10	11	12	13	14	Total
Broiler	66.5	66.3	66.6	67.0	67.1	66.6
	±0.7	±0.5	±1.0	±1.0	±1.0	±0.3
	(N=21)	(N=30)	(N=12)	(N=12)	(N=16)	(N=91)
White	66.5	65.7	67.2	66.2	66.3	66.3
Leghorn	±0.8	±0.7	±0.8	±1.3	±1.1	±0.3
	(N=21)	(N=30)	(N=12)	(N=12)	(N=16)	(N=91)

Numeric figure in the parentheses indicates sample size.

on Day 11 had HR decelerations of more than one during the 1-hour recording and the remaining five embryos had no HR decelerations. On Day 12, HR decelerations appeared in all ten embryos. HR decelerations increased in magnitude and appearing frequency in 13- and 14-day-old embryos and PSD markedly increased.

For White Leghorn chicken embryos, examples of 1-hour recording of IHR on individual incubation days and PSD are shown in Figure 2. The trend of development of HR decelerations was similar. IHR baseline was flat in a 10-day-old embryo and the remaining nine embryos on Day 10. In an 11-day-old embryo, five small decelerations appeared. On Day 11, HR decelerations of more than one appeared in six other embryos and the remaining three embryos had no IHR decelerations. On Day 12, all ten embryos had HR decelerations. HR decelerations increased in magnitude and appearing frequency on Days 13 and 14, and PSD markedly increased.

SPECTRAL POWER

Figure 3 shows SP of the frequency in LFR and HFR and the total SP in embryos of the broiler and White Leghorn chickens. In LFR (top panel), the difference between the broiler and White Leghorn chickens was not significant (P=0.215), but the differences between individual incubation days were significant (P<0.001). Tukey post hoc test indicates that the difference between Day 10 and Day 11 and the difference between Day 11 and Day 12 were insignificant, but the differences between the remaining days; that is, between Day 10 and Day 12 (P<0.05) onward (P<0.01), between Day 11 and Day 13 onward (P<0.01), between Day 12 and Day 13 (P<0.05) onward (P<0.01) and between Day 13 and Day 14 (P<0.01), were significant. In HFR (middle panel), the difference between the broiler and

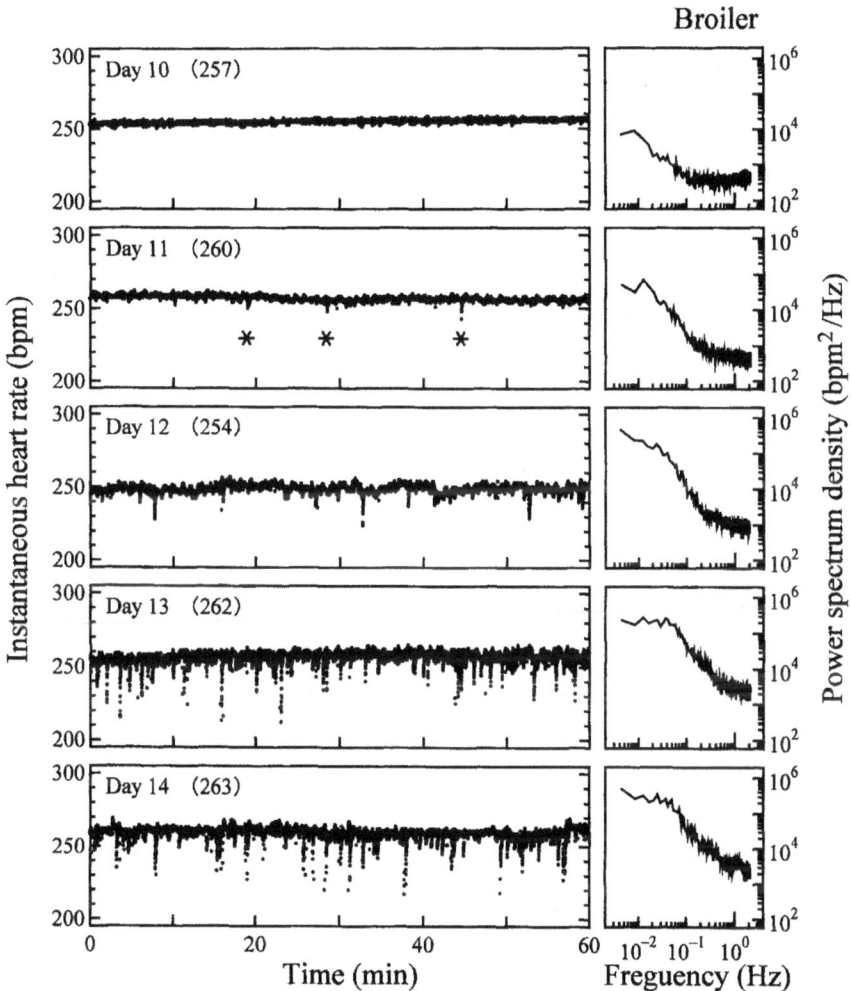

Figure 1. One-hour recordings of instantaneous heart rate (IHR) (left) and power spectrum density (right) on Days 10, 11, 12, 13 and 14 of incubation in broiler chicken embryos. Each point indicates individual IHR value. The numerical figure in the parentheses shows a value of mode heart rate (MHR) for 1-hour recording of IHR. An asterisk in the second panel indicates HR decrease which was judged as HR deceleration.

White Leghorn chickens was not significant (P=0.984), but the differences between individual incubation days were significant (P<0.001). Tukey post hoc test indicates that the differences between Days 10, 11 and 12 and the difference between Day 13 and Day 14 were insignificant (P>0.05), but the differences between remaining days were significant. For total SP (bottom panel), the difference between broiler and White Leghorn chickens was not significant (P=0.407), but the differences between individual incubation days were significant (P<0.001). Tukey post hoc test indicates that the differences between Day 10 and Day 11, the difference between Day 11 and Day 12 and the difference between Day 13 and Day 14 were insignificant (P>0.05), but the differences between the remaining days were significant.

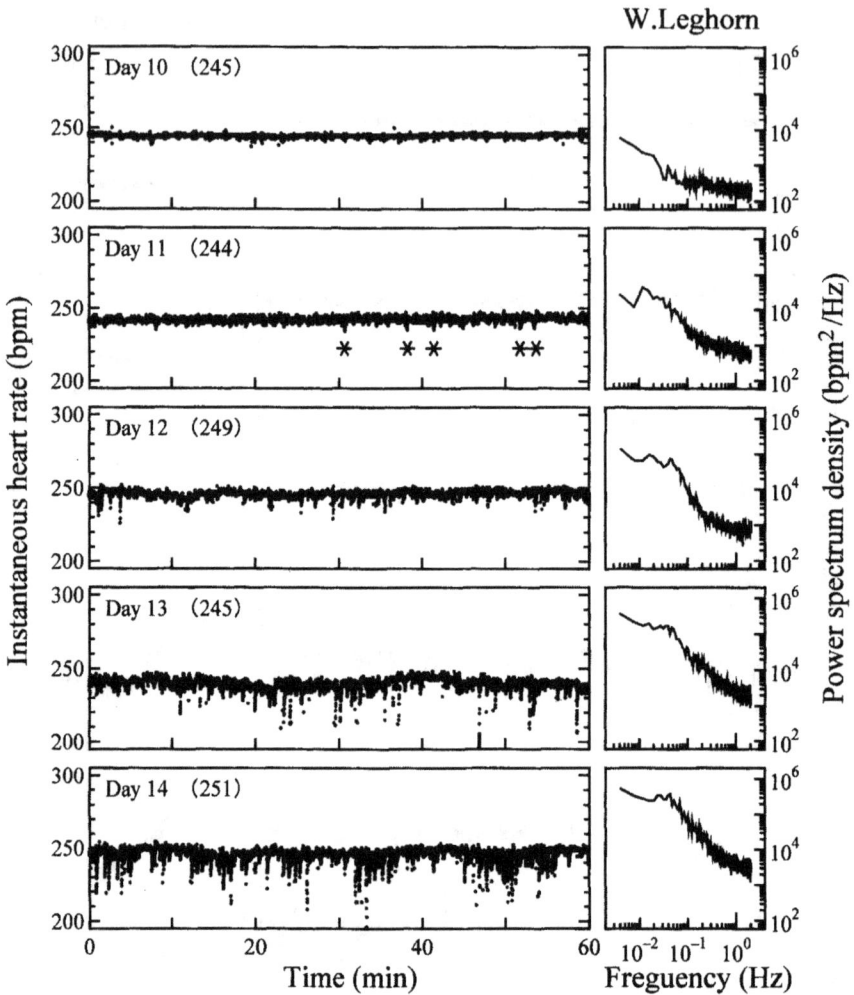

Figure 2. One-hour recordings of instantaneous heart rate (IHR) and power spectrum density on Days 10, 11, 12, 13 and 14 of incubation in White Leghorn chicken embryos. Other symbols and statements are the same as in Figure 1.

MODE HEART RATE

Figure 4 presents daily changes (i.e., developmental pattern) of embryonic MHR in the broiler and White Leghorn chickens. The difference between the broiler and White Leghorn chicken embryos was significant ($P<0.0001$) and the differences between individual incubation days were also significant ($P=0.003$). Tukey post hoc test indicates that the differences between Days 10, 11, 12 and 13 and the difference between Day 13 and Day 14 were not significant, but the differences between Day 10 and Day 14 ($P<0.01$), between Day 11 and Day 14, and between Day 12 and Day 14 ($P<0.05$) were significant.

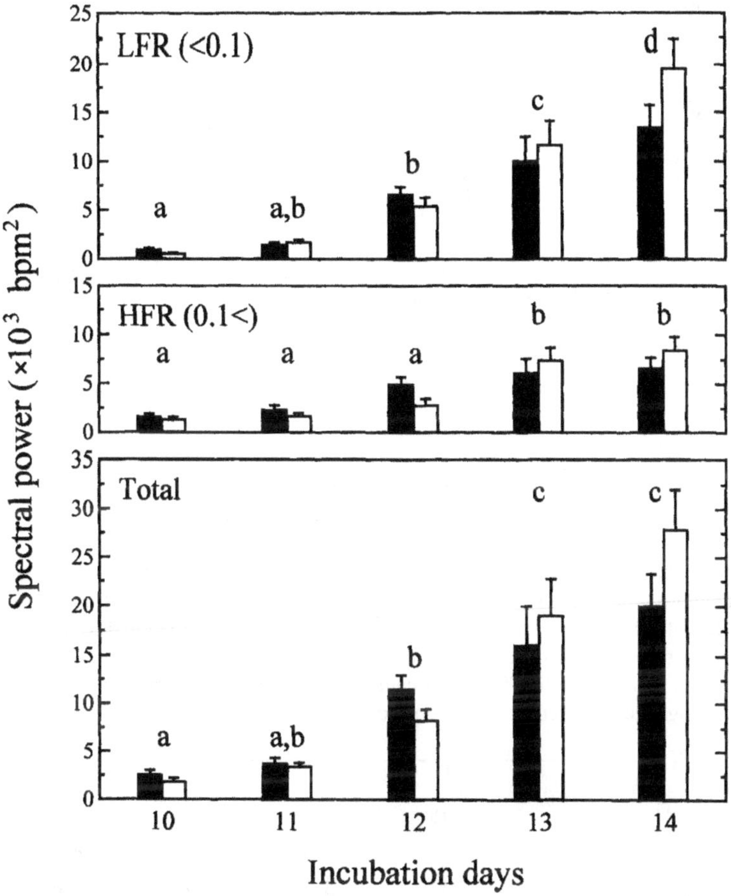

Figure 3. Spectral power (SP) of frequency in low frequency region (LFR) of less than 0.1 Hz (top panel) and high frequency region (HFR) of more than 0.1 Hz (middle panel), and total SP (bottom panel) in broiler (shaded square) and White Leghorn (open square) chicken embryos. Vertical bar indicates standard error. The same letters indicate that the difference between group means is not significant, and significant difference between group means is shown by different letters.

EGGSHELL CONDUCTANCE

Figure 5 presents eggshell GO_2 of the broiler and White Leghorn chickens approximated from the water loss during 10, 11, 12, 13 and 14 days of incubation. GO_2 of White Leghorn chicken eggs was significantly higher than that of broiler chicken eggs (P<0.0001), but GO_2 determined on any days of incubation from Day 10 to Day 14 was not different between incubation days (P=0.102). The mean value of GO_2 of 91 eggs was 15.9±0.3 and 17.7±0.3 ml•day^{-1}•torr^{-1} for the broiler and White Leghorn chicken eggs, respectively. The egg mass-specific conductance (gO_2) was 0.239±0.004 and 0.268±0.004 ml•day^{-1}•torr^{-1}•g^{-1} for the broiler and White Leghorn chicken eggs, respectively.

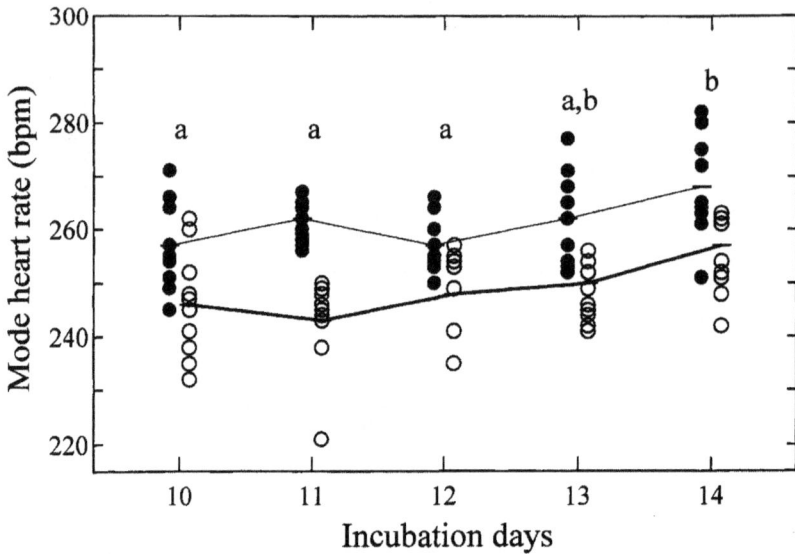

Figure 4. Daily changes (developmental pattern) in mode heart rate (MHR) in broiler (closed circles) and White Leghorn (open circles) chicken embryos. The number of eggs determined for MHR was 10 for the broiler and White Leghorn chickens, respectively, on each incubation day. Mean value on the individual incubation days are connected by thin and thick solid lines for broiler and White Leghorn chickens. The same letters indicate insignificant difference between group means, and the different letters indicate significant difference between group means.

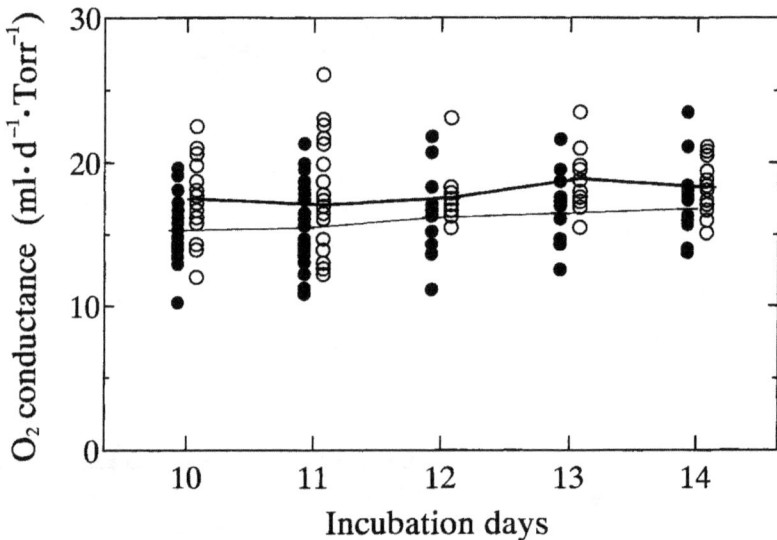

Figure 5. The eggshell oxygen conductance (GO_2) estimated from the water loss of eggs on Days 10, 11, 12, 13 and 14 of incubation in the broiler (closed circles) and White Leghorn (open circles) chickens. Equal number of eggs of broiler and White Leghorn chickens were determined for GO_2 on each incubation day; that is, 21, 30, 12, 12 and 16 eggs for Days 10, 11, 12, 13 and 14, respectively. Mean value on each incubation day is connected by thin and thick solid lines for broiler and White Leghorn chickens.

Discussion

No significant difference of egg mass between different strains of chicken is a favourable requirement for comparison of embryonic physiological variables. Embryo size is correlated with egg mass (Jull and Heywang, 1930; Bray and Iton, 1962; Xu and Mortola, 1988). If the egg mass is different, the respiratory surface area of the eggshell becomes different because of the dependence of shell surface area on the egg mass (Besch *et al.*, 1968; Paganelli *et al.*, 1974). It is possible that the different respiratory surface area of the eggshell alters diffusive gas exchange of embryos due to different eggshell gas conductance and influences growth of embryos (Xu and Mortola, 1988). Accordingly, development of embryonic physiological variables during incubation may be different between small and large embryos. Although these matters are beyond the scope of the present report, selection of identical egg mass between the two strains avoids a possible influence on embryonic HR due to different gas exchange and different growth of embryos.

In human foetuses, HR stable to within ±10 pbm for more than 80% of the recording time was referred to as the baseline heart rate (Wheeler and Murrills, 1978). Thus, because the baseline is not always constant and has some variation, it is difficult to determine visually HR deceleration when a drop of IHR is small. For instance, the first two HR drops in the second panel of Figure 1 seem to be deceleration, but they did not deviate from HR baseline more than 10 bpm and are rid of HR deceleration. Similarly, several HR drops in the second panel of Figure 2 are not HR decelerations. Accordingly, appearing frequency of IHR decelerations is not always determined visually when HR drops are small. However, in both strains in the present experiment, it is seen that IHR baseline did not suddenly drop in all 10-day-old embryos and clear IHR decelerations appeared in all 12-day-old embryos. On day 11, IHR baseline was flat in some embryos and obvious HR decelerations appeared in some other embryos. Development of IHR decelerations appeared to be identical in both strains.

For possible quantitative comparison, SP was calculated for individual embryos. In the previous studies, we showed that IHR accelerations appeared in addition to decelerations in late stage embryos (Höchel *et al.*, 1998; Chiba *et al.*, 2004). Figure 6 presents examples of 1-hour recording of IHR measured from two 18-day-old embryos and PSD. Both embryos were given 20 µg-atropine through a venous catheter at time of 30 min. IHR decelerations were eliminated by atropine administration, but IHR accelerations remained. PSD calculated for the first 30-min recording is shown by a black curve and PSD for the last 30-min recording having no HR decelerations is shown by a grey curve. Apparently, PSD in HFR was decreased in IHR recordings having no HR decelerations, indicating that HR decelerations have high power mainly in HFR. In fact, the previous study suggested that HR decelerations had high power in HFR and HR accelerations had high power in LHR (Chiba *et al.*, 2004).

Development of SP calculated from PSD in HFR was not different between the broiler and White Leghorn chicken embryos (Figure 3). SP in LFR and total SP were also not different between both strains. As a result, it may be concluded that development of IHR decelerations in embryos is not different between the broiler and White Leghorn chickens. In both strains, IHR decelerations appeared in all the embryos on Day 12 and increased in magnitude and appearing frequency on the following days. SP in HFR on Day 12 was not significantly different from that on Days 10 and 11. As far as SP was concerned, HR decelerations on Day 12 were not detected by power spectrum analysis. Meanwhile, total SP was different between Day 10 and Day 12, and between Day 12 and Days 13 and 14.

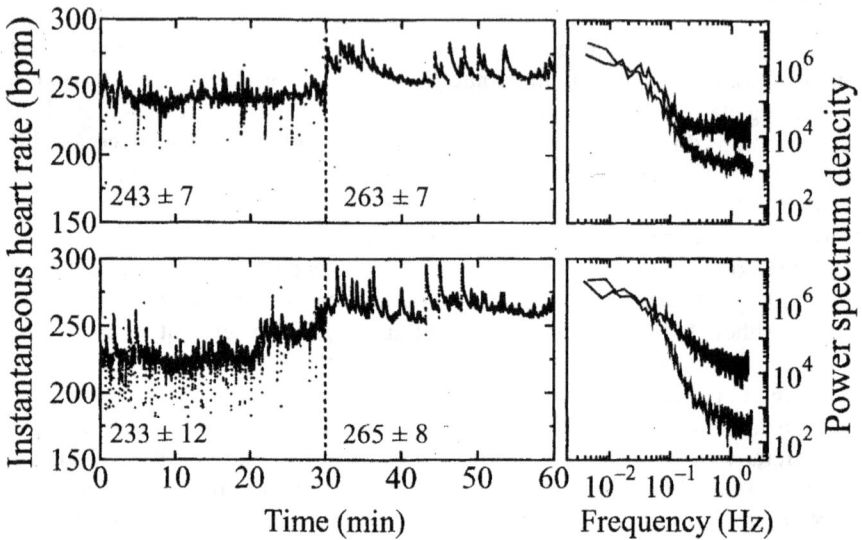

Figure 6. Examples of instantaneous heart rate (IHR) and power spectrum density before and after administration of atropine, previously determined for two 18-day-old embryos of broiler chicken. Atropine was given at 30 min shown by dotted line

This statistical analysis seems consistent with visual estimation of development of HR decelerations.

The simultaneous incubation and measurement of embryonic HR in the broiler and White Leghorn chickens excluded adverse effects of difference of incubation temperature, incubation time, measuring environments and methodology on embryonic HR. In addition, egg mass was statistically the same between both strains. MHR was the value calculated from continuous recording of IHR for 1-hour period. With these careful arrangements, we anticipated no significant difference of MHR between both strains. A previous study also reported that no significant difference was observed in the average daily HR of chicken embryos from four different genotypes (Cain *et al.*, 1967), although sex difference was reported on embryonic HR (Cain *et al.*, 1967; Laughlin *et al.*, 1976; Glahn *et al.*, 1986). However, developmental pattern of MHR was significantly different between the two strains in the present experiment and MHR was consistently higher in broiler than White Leghorn chicken embryos.

At the Workshop, we had the suggestion that genetic difference of body temperature (Tb) between the broiler and layer chicken embryos should be taken into consideration for the difference of HR and in addition the eggshell conductance might influence the HR difference. Previous studies have shown that during Days 12 to 20 of incubation, embryonic heat production in White Leghorn chickens was about 26-30% lower than that of two broiler lines, although the egg mass of the layer was 6-8% lighter than that of the broilers (Janke *et al.*, 2004, 2005). Concurrently, embryonic Tb of the White Leghorn chickens was 0.5-0.6°C lower than that of the broilers. Temperature coefficient of HR in 12-day-old chicken embryos exposed to high or low ambient temperature for 5 hours was Q_{10}=2 or slightly higher than 2 (Tazawa *et al.*, 1992; Ono *et al.*, 1994). Assuming that Q_{10} of embryonic HR in both strains is 2, Tb difference of 0.5-0.6°C causes HR difference of 8-10 bpm for HR of 260 bpm. These values are close to the average HR difference of about 13 bpm between both strains (Fig. 4). It may be possible that high embryonic HR of the broiler compared with the White Leghorn reflects the genetic difference of embryonic Tb between both strains.

The eggshell GO_2 or GH_2O approximated from the water loss of eggs during incubation in the present report was in the range previously determined by using the calibrated egg technique (Tullett and Deeming, 1982; Burton and Tullett, 1983, 1985; Tazawa *et al.*, 1983, 1988; Visschedijk, A.H.J., *et al.*, 1985; Okuda and Tazawa, 1988; Xu and Mortola, 1988; Burton *et al.*, 1989). The eggshell conductance and mass-specific conductance as well as the egg mass widely varied among fertilized chicken eggs in a commercial hatchery (Tullett and Deeming, 1982; Visschedijk *et al.*, 1985). Two hundred eggs of a broiler strain of which mean GH_2O was 17.34 mg•mmHg^{-1}•day^{-1} spanned GH_2O from 9.24 to 24.35 mg•mmHg^{-1}•day^{-1} which were equivalent to mass-specific GH_2O (i.e., gH_2O) of 0.130 to 0.399 mg•mmHg^{-1}•day^{-1}•g^{-1} (Tullett and Deeming, 1982). GH_2O of 395 layer strain eggs ranged from 7.1 to 32.7 mg•day^{-1}•torr^{-1} with mean of 15.5 mg•day^{-1}•torr^{-1}, corresponding to gH_2O of 0.122 to 0.479 mg•day^{-1}•torr^{-1}•g^{-1} with mean of 0.246 mg•day^{-1}•torr^{-1}•g^{-1} (Visschedijk *et al.*, 1985). The wide variability of shell conductance affected the growth, gas exchange, acid-base balance and other respiratory variables in chicken embryos (Tullett and Deeming, 1982; Burton and Tullett, 1983, 1985; Tazawa *et al.*, 1983, 1988; Visschedijk, A.H.J., *et al.*, 1985; Okuda and Tazawa, 1988; Xu and Mortola, 1988; Burton *et al.*, 1989).

The present measurement showed that GO_2 ranged from 10.5 to 23.5 ml•day^{-1}•torr^{-1} (gO_2; 0.163 to 0.365 ml•day^{-1}•torr^{-1}•g^{-1}) for 91 broiler chicken eggs and from 12.2 to 26.1 ml•day^{-1}•torr^{-1} (gO_2; 0.182 to 0.400 ml•day^{-1}•torr^{-1}•g^{-1}) for 91 White Leghorn chicken eggs (Figure 5). The mean value was significantly different between the two strains. The eggshell conductance was about 10% lower in the broiler than in the White Leghorn chickens. Judging from the previous reports on the relation between gas exchange and eggshell gas conductance (Tazawa *et al.*, 1983; Visschedijk *et al.*, 1985; Okuda and Tazawa, 1988), 10% deviation of conductance from the mean value will not affect gas exchange and blood gas variables in chicken embryos. For eggs of which shell conductance was artificially, widely altered, oxygen uptake and arterialized blood Po_2 of 16-day-old embryos were shown as a function of shell conductance (Okuda and Tazawa, 1988). GO_2 of 15.9 and 17.7 ml•day^{-1}•torr^{-1} yielded oxygen uptake of 492 and 496 ml•day^{-1}, respectively. $gO2$ of 0.239 and 0.268 ml•day^{-1}•torr^{-1}•g^{-1} yielded Po_2 of 55 and 57 mmHg. Similarly, for 16-day-old chicken embryos with naturally varying eggshell conductance, an equation expressing the relation between blood Po_2 and shell conductance was reported (Tazawa *et al.*, 1983). The equation yielded blood Po_2 difference of 2 mmHg between eggs with $gO2$ of 0.239 and 0.268 ml•day^{-1}•torr^{-1}•g^{-1}. As a result, about 10% difference of shell conductance caused only small differences of variables. Accordingly, it is unlikely that significant difference in embryonic MHR between the broiler and White Leghorn chickens was attributed to the difference of eggshell conductance.

In the two separate reports on eggshell conductance in commercial hatcheries (Tullett and Deeming, 1982; Visschedijk *et al.*, 1985), mean GH_2O was larger in the broiler than in the layer strain eggs, although the statistical significance was unknown. In the present simultaneous measurement, the eggshell conductance of the White Leghorn chicken eggs was statistically larger than that of the broiler chicken eggs. It is not known whether the significant difference was due to genetic, environmental or maternal influences caused by environmental experiences of the mother (Bernardo, 1996; Burggren, 1999). Further simultaneous measurements with various strains are needed.

Acknowledgement

We are grateful to Ms. K. Kunishige, Hokkaido Animal Research Center, Shintoku, Hokkaido, for providing us with fertile eggs of White Leghorn chickens.

References

AUBERT, A.E., BECKERS, F., RAMAEKERS, D., VERHEYDEN, B., LERIBAUX, C., AERTS, J-M and BERCKMANS, D. (2004) Heart rate and heart rate variability in chicken embryos at the end of incubation. *Experimental Physiology* **89**: 199-208.

BERNARDO, J. (1996) Maternal effects in animal ecology. *American Zoology* **36**: 83-105.

BESCH, E.L., SLUKA, S.J. and SMITH, A.H. (1968) Determination of surface area using profile recordings. *Poultry Science* **47**: 82-85.

BRAY, D. F. and ITON, E.L. (1962) The effect of egg weight on strain differences in embryonic and post embryonic growth in the domestic fowl. *British Poultry Science* **3**: 175-186.

BURGGREN, W.W. (1999) Genetic, environmental and maternal influences on embryonic cardiac rhythms. *Comparative Biochemistry and Physiology* A **124**: 423-427.

BURTON, F.G. and TULLETT, S.G. (1983) A comparison of the effects of eggshell porosity on the respiration and growth of domestic fowl, duck and turkey embryos. *Comparative Biochemistry and Physiology* **75A**: 167-174.

BURTON, F.G. and TULLETT, S.G. (1985) The effects of egg weight and shell porosity on the growth and water balance of the chicken embryo. *Comparative Biochemistry and Physiology* **81A**: 377-385.

BURTON, F.G., STEVENSON, J.M. and TULLETT, S.G. (1989) The relationship between eggshell porosity and air space gas tensions measured before and during the parafoetal period and their effects on the hatching process in the domestic fowl. *Respiration Physiology* **77**: 89-100.

CAIN, J.R., ABBOTT, U.K. and ROGALLO, V.L. (1967) Heart rate of the developing chick embryo. *Proceedings for Society of Experimental Biology and Medicine* **126**: 507-510.

CHIBA, Y., FUKUOKA, S., NIIYA, A., AKIYAMA, R. and TAZAWA, H. (2004) Development of cholinergic chronotropic control in chick (*Gallus gallus domesticus*) embryos. *Comparative Biochemistry and Physiology* A **137**: 65-73.

CROSSLEY II, D. and ALTIMIRAS, J. (2000) Ontogeny of cholinergic and adrenergic cardiovascular regulation in the domestic chicken (*Gallus gallus*). *American Journal of Physiology (Reg. Integr. Comp. Physiol.)* **279**: R1091-R1098.

CROSSLEY II, D., BURGGREN, W.W. and ALTIMIRAS, J. (2002) Cardiovascular regulation during hypoxia in embryos of the domestic chicken *Gallus Gallus. American Journal of Physiology (Reg. Integr, Comp. Physiol.)* **284**: R219-R226.

CROSSLEY II, D.A., HICKS, J.W. and THORNBURG, K. (2003) A comparison of cardiovascular regulatory mechanisms present during fowl ontogeny; White Leghorn vs. bantam chickens. Abstr. SICB 147.

GLAHN, R.P., MITSOS, W.J. and WIDEMAN, R.F.JR. (1987) Evaluation of sex differences in embryonic heart rates. *Poultry Science* **66**: 1398-1401.

HÖCHEL, J., AKIYAMA, R., MASUKO, T., PEARSON, J.T., NICHELMANN, M. and TAZAWA, H. (1988) Development of heart rate irregularities in chick embryos.

American Journal of Physiology 275 (*Heart Circ. Physiol.*) **44**: H527-H533.

HOYT, D.F., BOARD, R.G., RAHN, H. and PAGANELLI, C.V. (1979) The eggs of the anatidae: Conductance, pore structure and metabolism. *Physiological Zoology* **52**: 438-450.

JANKE, O., TZSCHENTKE, B. and BOERJAN, M. (2004) Comparative investigation of heat production and body temperature in embryos of modern chicken breeds. *Avian and Poultry Biology Reviews* **15**: 191-196.

JANKE, O., KEKLU, Y., BOERJAN, M. and TZSCHENTKE, B. (2005) Development of heat production and body temperature in precocial bird embryos – A comparative analysis. *Abstract of the 2nd Combined Workshop Fundamental* Physiology of European Working Group of Physiology and Perinatal Development in *Poultry.* p.26, Berlin, Sept. 23-25.

JULL, M.A. and HEYWANG, B.W. (1930) Yolk assimilation during the embryonic development of the chick. *Poultry Science* **9**: 393-404.

KHANDOKER, A.H., DZIALOWSKI, E.M., BURGGREN, W.W. and TAZAWA, H. (2003) Cardiac rhythms of late pre-pipped and pipped chick embryos exposed to altered oxygen environments. *Comparative Biochemistry and Physiology* A **136**: 289-299.

KHANDOKER, A.H., FUKAZAWA, K., DZIALOWSKI, E.M., BURGGREN, W.W. and TAZAWA, H. (2003) Maturation of the homeothermic response of heart rate to altered ambient temperature in developing chick hatchlings (*Gallus gallus domesticus*). *American Journal of Physiology Regul. Integr. Comp. Physiol.* **286**; R129-R137.

LAUGHLIN, K.F., LUNDY, H. and TAIT, J.A. (1976) Chick embryo heart rate during the last week of incubation: population studies. *British Poultry Science* **17**: 293-301.

MORIYA, K., HÖCHEL, J., PEARSON, J.T. and TAZAWA, H. (1999) Cardiac rhythms in developing chicks. *Comparative Biochemistry and Physiology* A **124**:461-468.

MORIYA, K., PEARSON, J.T., BURGGREN, W.W., AR, A. and TAZAWA, H. (2000) Continuous measurements of instantaneous heart rate and its fluctuations before and after hatching in chickens. *Journal of Experimental Biology* **203**: 895-903.

OKUDA, A. and TAZAWA, H. (1988) Gas exchange and development of chicken embryos with widely altered shell conductance from the beginning of incubation. *Respiration Physiology* **74**: 187-198.

ONO, H., HOU, P.-C.L. and TAZAWA, H. (1994) Responses of developing chicken embryos to acute changes in ambient temperature: noninvasive study of heart rate. *Israel Journal of Zoology* **40**: 467-479.

PAGANELLI, C.V., OLSZOWKA, A. and AR, A. (1974) The avian egg: surface area, volume, and density. *The Condor* **76**: 319-325.

SOROKIN, Y., DIERKER, L.J., PILLAY, S.K., ZADOR, I.E., SCHREINER, M.L. and ROSEN, M.G. (1982) The association between fetal heart rate patterns and fetal movements in pregnancies between 20 and 30 weeks' gestation. *American Journal of Obstetrics and Gynecology* **143**: 243-249.

TAZAWA, H., VISSCHEDIJK, A.H.J. and PIIPER, J. (1983) Blood gases and acid-base status in chicken embryos with naturally varying egg shell conductance. *Respiration Physiology* **54**: 137-144.

TAZAWA, H., NAKAZAWA, S., OKUDA, A. and WHITTOW, G.C. (1988) Short-term effects of altered shell conductance on oxygen uptake and hematological variables of late chicken embryos. *Respiration Physiology* **74**: 199-210.

TAZAWA, H., YAMAGUCHI, S., YAMADA, M. and DOI, K. (1992) Embryonic heart rate of the domestic fowl (*Gallus domesticus*) in a quasiequilibrium state of altered ambient temperatures. *Comparative Biochemistry and Physiology* **101A**: 103-108.

TULLETT, S.G. (1984) The porosity of avian eggshells. *Comparative Biochemistry and Physiology* **78A**: 5-13.

TULLETT, S.G. and DEEMING, D.C. (1982) The relationship between eggshell porosity and oxygen consumption of the embryo in the domestic fowl. *Comparative Biochemistry and Physiology* **72A**: 529-533.

VISSCHEDIJK, A.H.J., TAZAWA, H. and PIIPER, J. (1985) Variability of shell conductance and gas exchange of chicken eggs. *Respiration Physiology* **59**: 339-345.

WHEELER, T. and MURRILLS, A. (1978) Patterns of fetal heart rate during normal pregnancy. *British Journal of Obstetrics and Gynaecology* **85**: 18-27.

XU, L. and MORTOLA, J.P. (1988) Development of the chick embryo: effects of egg mass. *Respiration Physiology* **74**: 177-186.

Does sequence of exposure to altered ambient temperatures affect the endothermic heart rate response of newly hatched chicks?

H. YONETA, K. FUKAZAWA, E.M. DZIALOWSKI[1], W.W. BURGGREN[1] AND H. TAZAWA*

Department of Electrical and Electronic Engineering, Muroran Institute of Technology, Muroran, 050-8585, Japan; [1]Department of Biological Sciences, P.O.Box 305220, University of North Texas, Denton, Texas 76203, USA

In chicken hatchlings (Days 0-7) that were brooded at a room temperature (24-27°C), development of homeothermic HR response was previously investigated while acutely exposing them first to ambient temperature (Ta) of 25°C and then 35°C with subsequent return to 25°C for every 1-hour period. HR response in this sequence of warming and cooling turned from thermo-conformity to homeothermic changes during the day of hatching (Day 0) (Khandoker *et al.*, 2004). However, because chicken hatchlings prefer a brooding temperature of 35°C to 25°C, it is possible that the sequence of warming and cooling affects the homeothermic HR response. The present experiment was designed to elucidate whether the HR response to altered Ta (ΔTa) is affected by the inverse order of temperature exposure; that is, first exposing hatchlings to 35°C and then 25°C with subsequent return to 35°C (sequence of cooling and warming). Chicken hatchlings were brooded at Ta of 35°C during Days 0-7 and measured for responses of HR and cloaca temperature to ΔTa with sequence of cooling and warming. The thermo-conformity response during the early period of Day 0 turned to the endothermic response during the last half of Day 0. Thus, sequence of temperature exposure did not affect evaluation of development of endothermic HR response, nor did brooding temperature affect development of thermoregulatory competence.

Keywords: chicken hatchling; chick; cloaca temperature; endothermic response; exposing sequence; instantaneous heart rate; thermo-conformity response; thermoregulatory competence

Introduction

In chickens, it is reported that changes in heart rate (HR) under acute hyperthermal and hypothermal stress reflect the thermoregulatory response (Wilson, 1948; Harrison and Biellier, 1969; Mitchell and Siegel, 1973; Darre and Harrison, 1979, 1987). The HR of developing avian embryos and hatchlings also responds to alteration of ambient tempera-

---

*Corresponding author: TAZAWA, Hiroshi
E-mail: tazawa@mmm.muroran-it.ac.jp

ture (Ta) in accordance with development of thermoregulatory competence (Tazawa and Nakagawa, 1985; Tazawa *et al.*, 1992, 2001, 2002; Ono *et al.*, 1994; Tamura *et al.*, 2003; Khandoker *et al.*, 2004). If the thermoregulatory competence is not yet developed or very small, HR baseline changes in conformity with altered Ta (ΔTa); that is, HR baseline increases or decreases in response to warming or cooling, respectively (i.e., thermo-conformity response). In contrast, animals provided with full-blown endothermy decrease or increase HR baseline against warming or cooling, respectively. In the previous experiment investigating development of homeothermic HR response, newly hatched broiler chicks (Days 0-7) that were brooded at room temperature (24-27°C) were exposed first to Ta of 25°C and then to 35°C for a 1-hour period with subsequent return to 25°C-environment (Khandoker *et al.*, 2004). Instantaneous heart rate (IHR) was continuously measured for 1-hour period each at Ta of 25°C, 35°C and 25°C. IHR baseline increased in response to 35°C-exposure with subsequent decrease during 25°C-exposure; that is, thermo-conformity response. However, the thermo-conformity response was limited during the early period on the day of hatching (Day 0). IHR baseline remained unchanged or inversely changed in response to ΔTa in advanced hatchlings on late Day 0 and early Day 1. From late Day 1 onward, IHR baseline decreased and then increased in response to 35°C and 25°C exposures, respectively; that is, inverse temperature dependent response (thermoregulatory or endothermic response). As a result, mean HR responses to changes in Ta of 10°C (ΔTa=±10°C) were reversed during the period from late Day 0 to early Day 1, showing progressive endothermic HR response.

Hatchlings are usually brooded at a warm temperature, so we hypothesized that the sequence of exposure starting from 25°C to 35°C and then 25°C (i.e., sequence of warming and cooling) may affect HR responses and accordingly alter the time course of development of endothermic response. In the present experiment, broiler chicken hatchlings were brooded at 35°C and their HR responses were measured in the order starting from 35°C to 25°C and then 35°C (sequence of cooling and warming). We examined whether the brooding temperature and the sequence of exposure affected the thermoregulatory competence and endothermic HR response, respectively. In addition, we measured cloaca temperature as a body temperature and compared this with skin temperature measured in the previous study.

Materials and methods

Fertile eggs of broiler chickens (*Gallus gallus domesticus*) were obtained from the same hatchery as in the previous experiment that determined the HR response to warming and cooling exposure (Khandoker *et al.*, 2004). The previous results are included in the present report for comparison as the data based on the exposure sequence of warming and cooling. In the present experiment, incubation and treatment of hatchlings were generally the same as the previous experiment. The time of hatching was determined by taking a digital photo every 30 min through the day and night. The individual chicks in a hatcher were identified by serial numbers indicating hatching time on a tag attached to the leg. They were then transferred to a brooding chamber warmed at 35°C and relative humidity of about 60% with food and water *ad libitum*. The brooding temperature of 35°C was different from that of 24-27°C used in the previous report (Khandoker *et al.* 2004). They were kept in the brooding chamber until the experiment was begun on Day 0 to Day 7.

IHR was determined in real time from an ECG with the aid of a computer as described by Moriya *et al.* (1999, 2000). Body temperature (Tb) was measured with a thin thermo-couple inserted into the cloaca as in a previous study (Tazawa *et al.*, 2004).

After placement of ECG electrodes and thermo-couple to chicks in the measuring cage, the cage was placed in a temperature-controlled chamber initially set at 35°C. The chicks were allowed to acclimate to the new circumstances for an hour, and Tb and IHR were recorded for the next 1 hour at 35°C. The temperature was then decreased to 25°C for another hour and increased again to 35°C for the final hour. The recordings of Tb and IHR were continued throughout this temperature exposure regime. Because the switch from thermo-conformity HR response to endothermic response occurred during Days 0 and 1 in the previous experiment (Khandoker *et al.*, 2004), newly hatched chicks on Days 0 and 1 were divided into the first 12-hour and the last 12-hour groups; that is, *group 0-11h* on Day 0, *group 12-23h* on Day 0, *group 24-35h* on Day 1 and *group 36-47h* on Day 1. Chicks were subjected once to the exposure test during the period from Day 0 to Day 7.

STATISTICAL ANALYSIS

One-way repeated measures ANOVA was used to examine the differences of Tb (or HR) between the consecutive three exposures to 35°C, 25°C and 35°C. When the ANOVA revealed significant differences, Tukey post hoc test was used to detect the differences of means between the three exposures. The differences of body mass between the previous group (Khandoker *et al.*, 2004; warming and cooling exposure test) and the present group (cooling and warming exposure) and between brooding days were examined by two-way factorial ANOVA with repeated measures and with Tukey post hoc test. The significant level was P<0.05. The mean values were shown with standard error (SE).

Results

MASS OF HATCHLINGS

Three-hour exposure experiments with the sequence of warming and cooling periods were previously made for 114 hatchlings that were brooded at Ta of 24-27°C with access to a heat lamp (Khandoker *et al.*, 2004). On Day 0, 61 hatchlings were examined in four horal groups; i.e., the first 8 hours, the second 8 hours, and two 4-hour groups during the last 8-hour period. Thus, the hatchlings in the second 8-hour group were divided into either the first 12-hour or the last 12-hour groups according to the time of present experiment. Thus, 19 and 42 hatchlings were regrouped into *group 0-11h* and *group 12-23h*, respectively. In the present experiment, 75 hatchlings were examined on Days 0 through 7 post hatch. The mass of hatchlings examined in both experiments is summarized in Table 1. The hatchlings examined on Days 4-7 were pooled in a single group, because their responses resulted in similar patterns. The mass of hatchlings was not statistically different between the two experiments (P=0.982). Meanwhile, the mass was significantly different between brooding days (P<0.0001); i.e., the advanced hatchlings on Days 4-7 were significantly heavier than hatchlings younger than 3 days.

RESPONSES TO THE SEQUENCE OF WARMING AND COOLING

In the previous experiment, skin temperature (Ts) was determined as a measure of peripheral Tb. Figure 1 shows responses of Ts and HR in four hatchlings on Days 0, 3 and 7. Two hatchlings on Day 0 are representative of early Day 0 group (panel A) and of late Day 0 group (panel B), respectively. Ts responded to the step-wise change in Ta from 25°C to 35°C with an exponential increase. In advanced hatchlings (panels C and D), the change in Ts was smaller. While the change in Ts decreased with growth of hatchlings, the response of HR baseline to warming turned from an increase in an early 0-day-old hatchling (panel A) to large decreases in advanced hatchlings (panels C and D) with in-between change in a late 0-day-old hatchling (panel B). For comparison, the mean value during the last 10-min period of individual exposures was calculated as the value of Ts (or mean heart rate; MHR). The difference of Ts (or MHR) between the two thermal environments was defined as the magnitude of change in Ts (or MHR); i.e., ΔTs (or ΔMHR).

Table 1. Mass (g) of hatchlings used in the previous warming and cooling experiment (Khandoker *et al.*, 2004; sequence 1) and in the present cooling and warming experiment (sequence 2).

Days	0		1		2	3	4~7
	(0-11h)	(12-23h)	(24-35h)	(36-47h)			
Sequence 1	48.6	47.5	47.5	49.1	48.9	51.3	58.3
	±0.9	±0.5	±0.5	±0.9	±0.3	±1.1	±2.4
	(N=19)	(42)	(13)	(11)	(10)	(7)	(12)
Sequence 2	49.5	48.4	48.5	48.0	51.2	51.5	53.8
	±0.8	±1.2	±1.8	±0.6	±1.5	±1.6	±0.9
	(12)	(11)	(9)	(11)	(9)	(9)	(14)

Figure 2 presents mean values of Ts and MHR during warming and cooling exposures on Days 0 to 7 post-hatch. Ts and MHR during the initial 25°C-exposure were defined as Ts_{ref} and MHR_{ref}, respectively. Ts and MHR during the 35°C-exposure were Ts_{35} and MHR_{35}, and those during the final 25°C-exposure were Ts_{25} and MHR_{25}. Ts was significantly different between Ts_{ref}, Ts_{35} and Ts_{25} at any ages (P<0.0001), and Ts_{ref} and Ts_{25} were different (P<0.01) in the youngest animals (*0-11h, 12-23h* and *24-35h*). MHR in *group 0-11h* changed in conformity with Ts and the difference in MHR between the three exposures; i.e., MHR_{ref}, MHR_{35} and MHR_{25}, was significant (P<0.0001). In advanced chicks on late Day 0 (*group 12-23h*), MHR_{35} increased significantly from MHR_{ref} (P<0.01) in conformity with Ts, and MHR_{25} further increased with final cooling (P<0.01). By early Day 1 (*group 24-35h*), the change in HR from MHR_{ref} to MHR_{35} was insignificant. In late Day 1 (*group 36-47h*) and more advanced hatchlings, HR response to altered Ta exhibited a mirror image of Ts.

RESPONSES TO THE SEQUENCE OF COOLING AND WARMING

Figure 3 presents examples of Tb and HR responses in four hatchlings on Days 0 and 1. In response to cooling and subsequent warming, Tb decreased and subsequently increased in an exponential fashion with different magnitudes between different age groups. As ΔTb decreased with the lapse of time, the HR response became endothermic. An early Day 0 chick decreased HR during cooling and increased it during warming (panel A). In an advanced Day 0 chick, HR baseline initially increased but reached a plateau during cooling (panel B). In Day 1 hatchlings, the plateau in HR during cooling was greater than HR at 35°C (panels C and D).

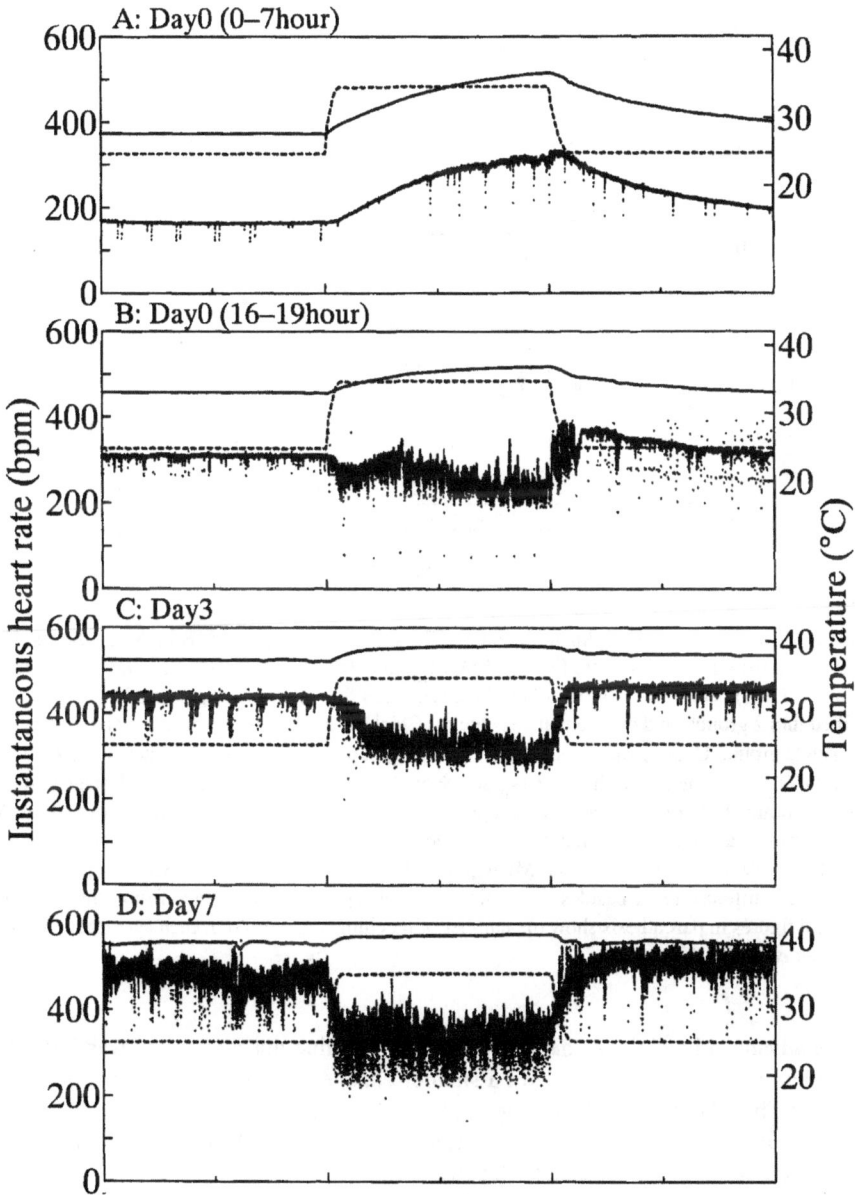

Figure 1. Responses of instantaneous heart rate (shown by individual points) and skin temperature (solid line) to altered ambient temperature (dashed line; 25°C-35°C-25°C) in four hatchlings on Days 0, 3 and 7. A; chick in *group 0-7h* which is regrouped into *group 0-11h* in this report, B; chick in *group 16-19h* (regrouped into *group 12-23h*), C; chick on Day 3, D; chick on Day 7.

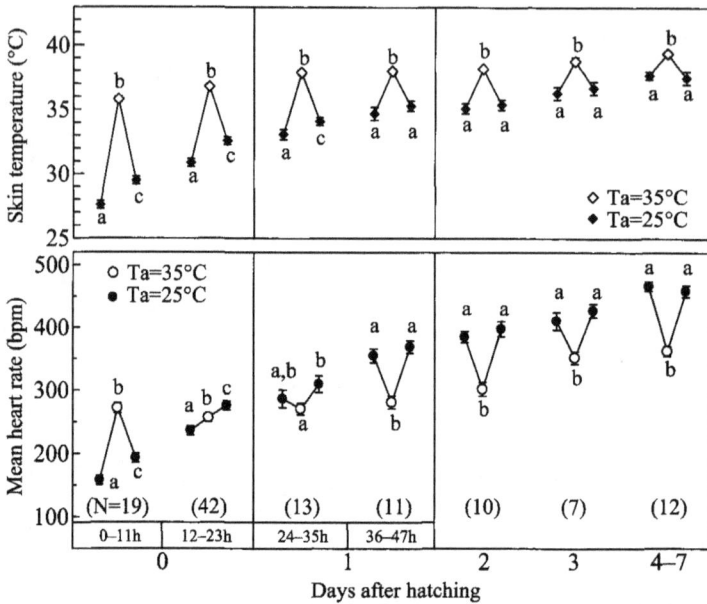

Figure 2. Mean value and SE of skin temperature (Ts) and mean heart rate (MHR) during the last 10-min period of exposure to the first 25°C (Ts$_{ref}$ and MHR$_{ref}$) and 35°C (Ts$_{35}$ and MHR$_{35}$) and the second 25°C (Ts$_{25}$ and MHR$_{25}$) determined for 61 0-day-old chicks classified into 2 horal groups, 24 1-day-old chicks classified into 2 groups, and other advanced chicks. Chicks examined on Days 4, 5, 6 and 7 were pooled into a single group. Upper panel: mean Ts of each age group connected by solid lines, with ◆ on the left indicating Ts$_{ref}$, ◇ in the centre showing Ts$_{35}$ and ◆ on the right showing Ts$_{25}$. For each stage of development, data points with different letters are significantly different from each other. Lower panel: MHR changes during warming and cooling with solid lines connecting MHR of each age group. ● on the left show MHR$_{ref}$, ○ in the centre indicate MHR$_{35}$, and ● on the right indicate MHR$_{25}$. For each stage of development, different letter besides data point indicates significant difference from other data points. Numerical figures in parenthesis show the number of hatchlings measured in each age group. Data from Khandoker *et al.*, 2004.

In more advanced hatchlings, the increase in HR baseline during cooling was further augmented and this was associated with a smaller change in ΔTb. Figure 4 presents the responses of Tb and HR of four hatchlings on Days 2, 3, 5 and 7. Tb$_{ref}$ exceeded 40°C in all the hatchlings, and ΔTb upon cooling decreased with growth of hatchlings. Contrarily, HR baseline during cooling increased with ages.

Figure 5 summarizes the responses of Tb and MHR to cooling and warming exposure in chicks on Days 0 to 7. Tb was significantly different between Tb$_{ref}$, Tb$_{25}$ and Tb$_{35}$ at any ages (P<0.0001) and the difference between Tb$_{ref}$ and Tb$_{35}$ was significant only in *group 0-11h*. In *group 0-11h*, MHR changed in conformity with Tb. The difference in MHR between the three exposures was significant (P<0.0001), but MHR$_{ref}$ and MHR$_{35}$ were not different. In *group 12-23h*, MHR was not different between the three environments (P=0.063). On early Day 1 (*group 24-35h*), the decrease from MHR$_{25}$ to MHR$_{35}$ became significant. In late Day 1 and more advanced chicks, the response of HR to ΔTa became the mirror image of ΔTb.

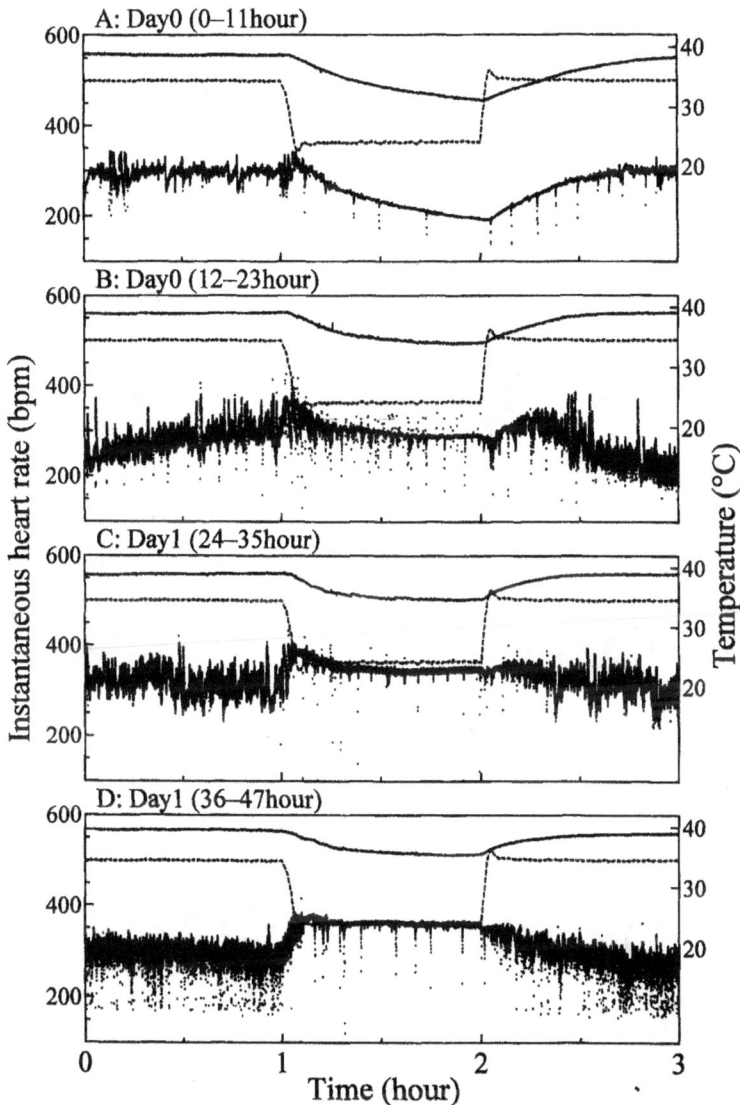

Figure 3. Responses of instantaneous heart rate (points) and cloaca temperature (Tb; solid line) to altered ambient temperature (dashed line; 35°C-25°C-35°C) in four hatchlings on Days 0 and 1. A; chick in *group 0-11h*, B; chick in *group 12-23h*, C; chick in *group 24-35h*, D; chick in *group 36-47h*. Tb and mean heart rate (MHR) during the last 10-min measurement in the first 35°C-exposure were defined as Tb_{ref} and MHR_{ref}, respectively. Tb and MHR during cooling were Tb_{25} and MHR_{25}, and those during the second 35°C-exposure were Tb_{35} and MHR_{35}. Tb was flat during the first 1-hour period at 35°C and exceeded more than 4°C over Ta in all hatchlings except early 0-day-old chick whose Tb was 38.7°C (A).

SKIN TEMPERATURE AND CLOACA TEMPERATURE

For comparison of Ts with Tb, both Ts and Tb during warming and cooling are plotted in Figure 6. The upper two curves depict Tb_{35} (open circle) and Ts_{35} (closed circle) at indi-

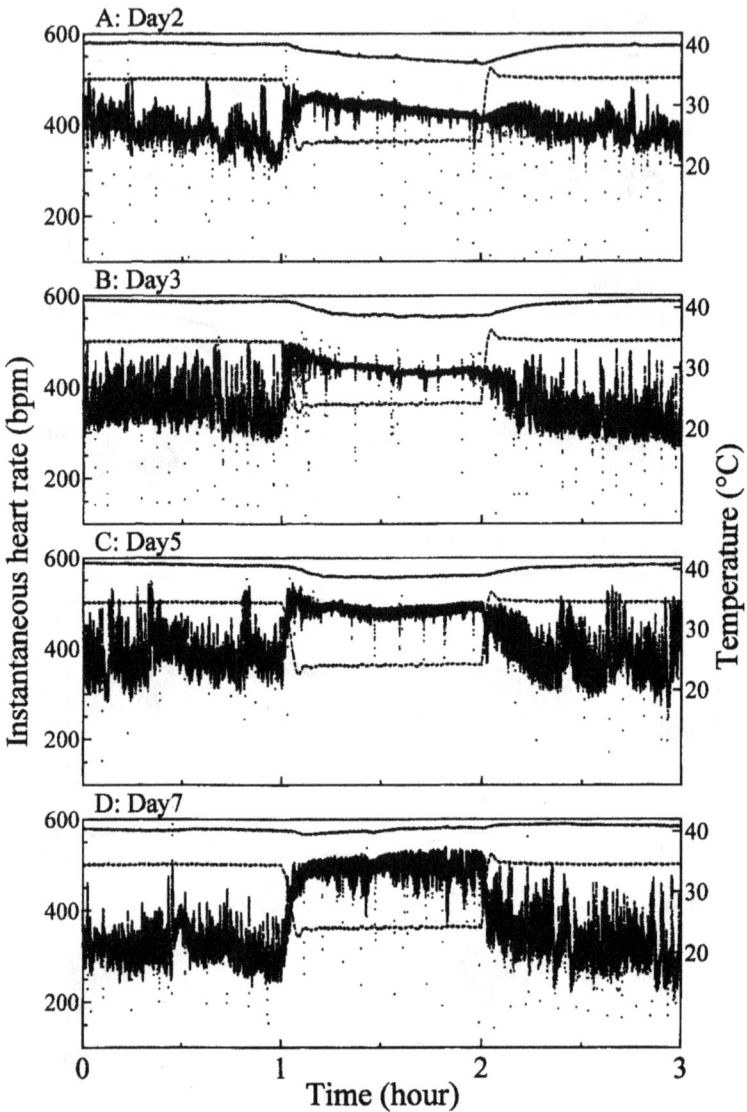

Figure 4. Responses of instantaneous heart rate (points) and cloaca temperature (solid line) to altered ambient temperature (dashed line; 35°C-25°C-35°C) in four hatchlings on Days 2-7. A; chick on Day 2, B; chick on Day 3, C; chick on Day 5, D; chick on Day 7. ΔTb changed from -3.1°C in a Day 2 chick to 0.3°C in a Day 7 chick. MHR_{25} increased from 416 bpm in the Day 2 chick to 492 bpm in the Day 7 chick.

vidual ages. Both the differences between Tb_{35} and Ts_{35} and the changes with development of hatchlings were significant (P<0.0001). Differences of temperature between Day 0 and Day 1 hatchlings were significant (P<0.01), but the difference between *group 24-35h* and *group 36-47h* was not significant. The temperature of Day 0 and Day 1 hatchlings differed significantly from advanced hatchlings on Day 2 onward (P<0.01). The temperatures of the

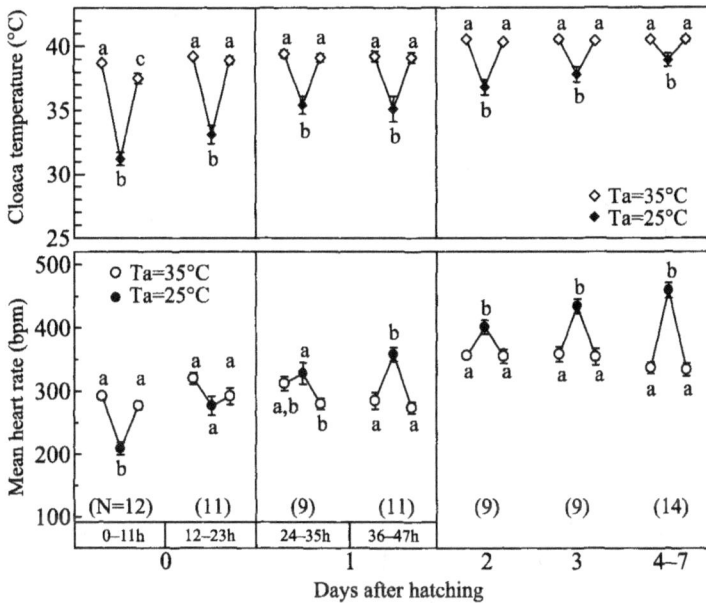

Figure 5. Mean value and SE of cloaca temperature (Tb) and mean heart rate (MHR) during the last 10-min period of exposure to the first 35°C exposure (Tb$_{ref}$ and MHR$_{ref}$), then 25°C (Tb$_{25}$ and MHR$_{25}$) and finally the second 35°C (Tb$_{35}$ and MHR$_{35}$), determined for 0-day-old chicks classified into 2 horal groups, 1-day-old chicks into 2 groups, and other advanced chicks. Chicks examined on Days 4, 5, 6 and 7 were pooled into a single group. Upper panel: mean Tb of each age group connected by solid lines, with ◇ on the left indicating Tb$_{ref}$, ◆ in the centre showing Tb$_{25}$ and ◇ on the right showing Tb$_{35}$. For each stage of development, data points with different letters are significantly different from each other. Lower panel: MHR changes during cooling and warming with solid lines connecting MHR of each age group. ○ on the left show MHR$_{ref}$, ● in the centre indicate MHR$_{25}$, and ○ on the right indicate MHR$_{35}$. For each stage of development, different letter besides data point indicates significant difference from other data points. Numerical figures in parenthesis show the number of hatchlings measured in each age group.

advanced hatchlings were not different. The lower two curves correspond to Tb$_{25}$ (open rhombus) and Ts$_{25}$ (closed rhombus). Both the difference between Tb$_{25}$ and Ts$_{25}$ and changes with development of hatchlings were significant (P<0.0001).

The difference between ΔTb and ΔTs was not significant (P=0.986), but changes with development of hatchlings were significant (P<0.0001). Tukey post hoc test indicates that two groups on Day 0 were significantly different from any other groups (P<0.01).

Discussion

Brooding temperature of the hatchlings was different between both the experiments. In the previous experiment, the hatchlings were brooded at Ta of 24-27°C until the experiment (Khandoker *et al.*, 2004). Then, the measurement of HR was begun with Ta of 25°C in order to avoid a large change in Ta which might impose thermal load on the hatchlings prior to the onset of measurement. In this context, the hatchlings in the present experiment were brooded

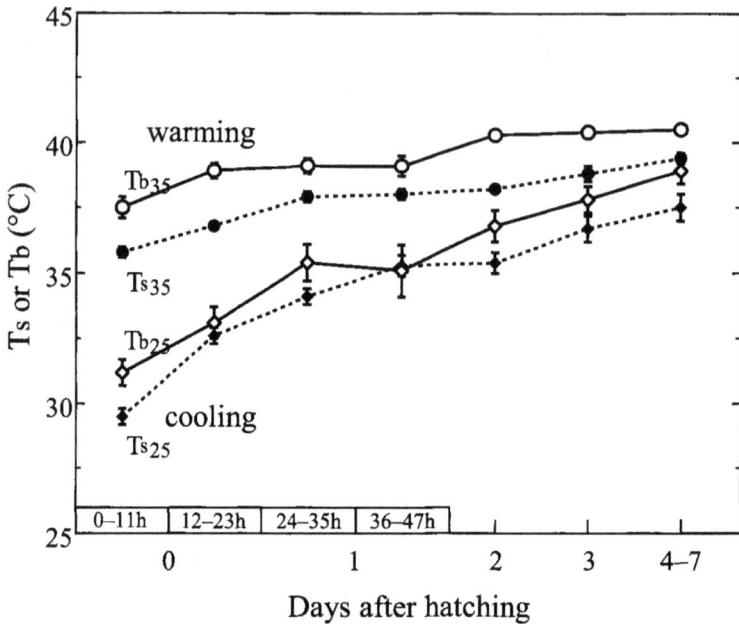

Figure 6. Mean value and SE of cloaca (Tb) and skin (Ts) temperatures during warming and cooling at each developmental stage connected by solid and broken lines, respectively. ○; Tb during warming (Tb$_{35}$), ●; Ts during warming (Ts$_{35}$), ◇; Tb during cooling (Tb$_{25}$) and ◆ ; Ts during cooling (Ts$_{25}$). Ts$_{35}$ and Ts$_{25}$ were on an average 1.6°C and 1.0°C lower than Tb$_{35}$ and Tb$_{25}$, respectively.

at Ta of 35°C from which cooling was begun. The mass of hatchlings was not different between the two experiments and thus difference in brooding temperature did not affect growth of hatchlings (Table 1).

In the experiment with warming and cooling sequence, Ts of the two Day 0 chicks increased exponentially (Figure 1), but ΔTs was apparently small in advanced chick (panel B) compared with early chick (panel A). ΔTs of Days 3 and 7 chicks further decreased (panels C and D). Apparently, thermoregulatory competence developed in advanced hatchlings more than early ones and the measurement of Ts indicates the development of thermoregulatory competence by the decrease in ΔTs. Meanwhile, HR responses reversed with growth of chicks. In the early Day 0 chick (panel A), HR baseline changed in parallel with the change in Ts and Ta (thermo-conformity response). On the same Day 0 (panel B), the advanced chick decreased HR during warming. The inverse change in HR against Ta (endothermic response) became pronounced with growth of hatchlings (panels C and D).

In the opposite sequence of exposure from 35°C to 25°C, Tb of Day 0 and Day 1 hatchlings decreased exponentially during cooling with different magnitudes (Figure 3). ΔTb was apparently small in Day 1 chicks (panels C and D) compared with Day 0 chicks (panels A and B). In advanced chicks more than Day 2, ΔTb decreased further, indicating further development of thermoregulatory competence with growth of chicks (Figure 4). The thermoregulatory competence of newly hatched chicks had also been indicated previously by endothermic metabolic response to exposure sequence of cooling and warming (Tazawa *et al.*, 2004). Meanwhile, HR response also changed in a similar time course to the experiment starting from warming. The change in HR response to the sequence of cooling and warming turned from thermo-conformity in the early Day 0 hatchling to endothermic pattern in advanced hatchlings (Figures 3 and 4).

Comparison of summarized results shown in Figure 5 with those in Figure 2 indicates that the endothermic change in HR response occurred over the same time course for both tests which were made with opposite sequence of exposure. On early Day 0, HR response was the thermo-conformity in both tests. On late Day 1 onward, HR response was endothermic and a mirror image in both tests. As a result, HR responses to ΔTa of 10°C turned from thermo-conformity to apparent endothermic pattern during the period from the last half of Day 0 to the first half of Day 1, while Tb (and Ts) deceased only the magnitude. In addition, it is concluded that the difference of brooding temperature by about 10°C did not change the development of endothermic HR response in chicks.

In the previous experiment, Ts was measured (Khandoker *et al.*, 2004). Although Ts and Tb were not measured simultaneously in the same animals and were not values at a steady-state, comparison of Ts with Tb indicates that Ts was significantly lower than Tb during both warming and cooling (Figure 6). However, comparison of ΔTs with ΔTb showed no difference between them, indicating that Ts is able to represent Tb in evaluating development of thermoregulatory competence. ΔTs or ΔTb markedly decreased during the first one and half days with gradual decrease onward. While ΔTs and ΔTb showed the development of thermoregulatory competence by means of a pattern of the decrease, HR measurement indicated the development of thermoregulatory competence in terms of the switch of the response from thermo-conformity to endothermic patterns.

The development of thermoregulatory competence in chicken embryos and hatchlings has been estimated by determination of metabolic response to altered Ta (Romijn and Lokhorst, 1951, 1955; Freeman, 1964, 1967, 1971; Wekstein and Zolman, 1969; Tazawa and Rahn, 1987; Tazawa *et al.*, 1988, 1989a, 1989b, 2004; Black and Burggren, 2004). In newly hatched chicks, metabolic response to ΔTa of 10°C was shown to be endothermic even during the first half of Day 0, indicating that chick hatchlings were more or less provided with the thermoregulatory competence (Tazawa *et al.*, 2004). Although the metabolic and HR responses were not determined simultaneously for the same hatchlings, it is likely that both responses are different in regard to the time course of the switch of response pattern. The switch to an endothermic response in HR suggests that autonomic nervous function develops during this period in relation to thermoregulatory competence. For instance, in newly hatched chicks, there was evidence that sympathetic blocker propranolol decreased partially the HR which was increased by cooling exposure, indicating that sympathetic nervous function is involved in development of thermoregulatory competence (Tazawa, 2005). Chickens which maintained their HR unchanged in a hot environment exhibited an increase in HR with atropine administration (Darre and Harrison, 1979), indicating that vagus nerve function regulates HR during hyperthermia. The autonomic nervous function remains to be studied as a possible mediator to regulate HR during development of thermoregulation in avian embryos and hatchlings.

In young and adult chickens, while an increase in HR in response to hyperthermia was reported (Whittow *et al.*, 1964), some studies showed that HR decreased or remained stable during exposure to a hot environment and increased in response to cold exposure as a result of thermoregulatory response of animals (Wilson, 1948; Harrison and Biellier, 1969; Mitchell and Siegel, 1973; Darre and Harrison, 1979, 1987). The cold- and heat-adapted chickens had high and low HR, respectively, compared with control chickens (Sturkie *et al.*, 1970).

The present study confirmed that these endothermic HR responses already occurred soon after hatching in chickens. In addition, it should be noted that in the emu the endothermic HR response already developed prior to hatching (Fukuoka *et al.*, 2006). Because it is likely that the development of endothermic HR response depends on a maturity of thermoregulatory competence in individual species of birds, further investigation into the endothermic HR response should be made for embryos and hatchlings in other domesticated birds for comparison. Either sequence of exposure to ΔTa may also be adopted for exposure test in other avian species.

Acknowledgments

This study was supported in part by National Science Foundation Grant No.IOB0417205 awarded to EMD and No.IBN0128043 to WWB and by a Grant-in-Aid for Scientific Research of the Japan Society for Promotion of Science (No. 15560352) to HT.

References

BLACK, J.L. and BURGGREN, W.W. (2004) Acclimation to hypothermic incubation in developing chicken embryo (*Gallus domesticus*) I. Developmental effects and chronic and acute metabolic adjustments. *Journal of Experimental Biology* **207**: 1543-1552.

DARRE, M.J. and HARRISON, P.C. (1979) Caloric value of cardiac response to hot environments. *Poultry Science* **58**: 807-809.

DARRE, M.J. and HARRISON, P.C. (1987) Heart rate, blood pressure, cardiac output, and total peripheral resistance of single comb white leghorn hens during an acute exposure to 35 C ambient temperature. *Poultry Science* **66**: 541-547.

FREEMAN, B.M. (1964) The emergence of the homeothermic-metabolic response in the fowl (*Gallus domesticus*). *Comparative Biochemistry and Physiology* **13**: 413-422.

FREEMAN, B.M. (1967) Some effects of cold on the metabolism of the fowl during the perinatal period. *Comparative Biochemistry and Physiology* **20**: 179-193.

FREEMAN, B.M. (1971) Body temperature and thermoregulation. In: *Physiology and Biochemistry of the Domestic Fowl.* (Ed. Bell, D.J. and Freeman, B.M.), Academic Press, New York, pp.1115-1151.

FUKUOKA, S., KHANDOKER, A.H., DZIALOWSKI, E.M., BURGGREN, W.W. and TAZAWA, H. (2006) Development of endothermic heart rate response in emu (*Dromaius novaehollandiae*) embryos. In: *New Insights into Fundamental Physiology and Perinatal Adaptation of Domestic Fowl.* (Ed. Yahav, S. and Tzschentke, B.), Nottingham University Press, Nottingham, pp. 29-42.

HARRISON, P.C. and BIELLIER, H.V. (1969) Physiological response of domestic fowl to abrupt changes of ambient air temperature. *Poultry Science* **48**: 1034-1044.

KHANDOKER, A.H., FUKAZAWA, K., DZIALOWSKI, E.M., BURGGREN, W.W. and TAZAWA, H. (2004) Maturation of the homeothermic response of heart rate to altered ambient temperature in developing chick hatchlings (*Gallus gallus domesticus*). *American Journal of Physiology: Regul. Integr. Comp. Physiol.* **286**: R129-R137.

MITCHELL, B.W. and SIEGEL, H.S. (1973) Physiological response of chickens to heat stress measured by radio telemetry. *Poultry Science* **52**: 1111-1119.

MORIYA, K., HÖCHEL, J., PEARSON, J.T. and TAZAWA, H. (1999) Cardiac rhythms in developing chicks. *Comparative Biochemistry and Physiology* 124A:461-468.

MORIYA, K., PEARSON, J.T., BURGGREN, W.W., AR, A. and TAZAWA. H. (2000) Continuous measurements of instantaneous heart rate and its fluctuations before and after hatching in chickens. *Journal of Experimental Biology* 203: 895-903.

ONO, H., HOU, P.-C.L. and TAZAWA, H. (1994) Responses of developing chicken embryos to acute changes in ambient temperature: noninvasive study of heart rate. *Israel Journal of Zoology* 40: 467-479.

ROMIJN, C. and LOKHORST, W. (1951) Foetal respiration in the hen. *Physiol. Comp. Oecol.* 2: 187-197.

ROMIJN, C. and LOKHORST, W. (1955) Chemical heat regulation in the chick embryo. *Poultry Science* 34: 649-654.

STURKIE, P.D., LIN, Y.-C. and OSSORIO, N. (1970) Effects of acclimatization to heat and cold on heart rate in chickens. *American Journal of Physiology* 219: 34-36.

TAMURA, A., AKIYAMA, R., CHIBA, Y., MORIYA, K., DZIALOWSKI, E.M. BURGGREN, W.W. and TAZAWA, H. (2003) Heart rate responses to cooling in emu hatchlings. *Comparative Biochemistry and Physiology* A 134: 829-838.

TAZAWA, H. and NAKAGAWA, S. (1985) Response of egg temperature, heart rate and blood pressure in the chick embryo to hypothermal stress. *Journal of Comparative Physiology* B 155: 195-200.

TAZAWA, H. and RAHN, H. (1987) Temperature and metabolism of chick embryos and hatchlings after prolonged cooling. *Journal of Experimental Zoology* **Suppl.** 1: 105-109.

TAZAWA, H., WAKAYAMA, H., TURNER, J.S. and PAGANELLI, C.V. (1988) Metabolic compensation for gradual cooling in developing chick embryos. *Comparative Biochemistry and Physiology* 89A: 125-129.

TAZAWA, H., OKUDA, A., NAKAZAWA, S. and WHITTOW, G.C. (1989a) Metabolic responses of chicken embryos to graded, prolonged alterations in ambient temperature. *Comparative Biochemistry and Physiology* 92A: 613-617.

TAZAWA, H., WHITTOW, G.C., TURNER, J.S. and PAGANELLI, C.V. (1989b) Metabolic responses to gradual cooling in chicken eggs treated with thiourea and oxygen. *Comparative Biochemistry and Physiology* 92A: 619-622.

TAZAWA, H., YAMAGUCHI, S., YAMADA, M. and DOI, K. (1992) Embryonic heart rate of the domestic fowl (*Gallus domesticus*) in a quasiequilibirum state of altered ambient temperatures. *Comparative Biochemistry and Physiology* 101A: 103-108.

TAZAWA, H., MORIYA, K., TAMURA, A., KOMORO, T. and AKIYAMA, R. (2001) Ontogenetic study of thermoregulation in birds. *Journal of Thermal Biology* 26: 281-286.

TAZAWA, H., MORIYA, K., TAMURA, A. and AKIYAMA, R. (2002) Low-frequency oscillation of instantaneous heart rate in newly hatched chicks. *Comparative Biochemistry and Physiology* A 131: 797-803.

TAZAWA, H., CHIBA, Y., KHANDOKER, A.H., DZIALOWSKI, E.M. and BURGGREN, W.W. (2004) Early development of thermoregulatory competence in chickens: responses of heart rate and oxygen uptake to altered ambient temperatures. *Avian and Poultry Biology Reviews* 15: 166-176.

TAZAWA, H. (2005) Cardiac rhythms in avian embryos and hatchlings. *Avian and Poultry Biology Reviews* 16: 123-150.

WEKSTEIN, D.R. and ZOLMAN, J.F. (1969) Ontogeny of heat production in chicks. *Federation Proceedings* **28**: 1023-1028.

WHITTOW, G.C., STURKIE, P.D. and STEIN, G. Jr. (1964) Cardiovascular changes associated with thermal polypnea in the chicken. *American Journal of Physiology* **207**: 1349-1353.

WILSON, W.O. (1948) Some effects of increasing environmental temperatures on pullets. *Poultry Science* **27**: 813-817.

Development of endothermic heart rate response in emu (*Dromaius novaehollandiae*) embryos

S. FUKUOKA, A.H. KHANDOKER, E.M. DZIALOWSKI[1], W.W. BURGGREN[1] AND H. TAZAWA*

Department of Electrical and Electronic Engineering, Muroran Institute of Technology, Muroran 050-8585, Japan; [1]Department of Biological Sciences, P.O.Box 305220, University of North Texas, Denton, Texas 76203, USA

Heart rate (HR) of avian embryos and hatchlings changes in response to an altered ambient temperature (Ta). In newly hatched chicks, it has been shown that the direction of change in HR upon cooling depends on the development of their thermoregulatory competence. In contrast, HR of newly hatched emu responds to cooling with an increase on the first day of hatching (Day 0). Accordingly, it is assumed that emu HR also exhibits a pattern of development of thermoregulatory competence as in the chicken, but the endothermic response of emu HR develops prior to hatching. In order to confirm the above assumption, emu embryos were exposed to Ta of 36°C, 26°C and 36°C for 90-min periods, respectively, during three distinct developmental stages; i.e., prior to internal pipping (pre-IP), internal pipping (IP) and external pipping (EP), and instantaneous heart rate (IHR) and egg temperature (Te) were measured. HR baseline of pre-IP embryos changed in the same direction as Ta, exhibiting thermo-conformity response. IP embryos initially responded to cooling with a transient increase in HR, but HR decreased as the cooling continued, exhibiting an incomplete endothermic response. On the contrary, most EP embryos markedly increased HR baseline in response to cooling, exhibiting endothermic thermoregulatory response. In the emu, endothermic HR response develops prior to hatching, providing EP embryos with a marked endothermic response of HR to altered Ta of 10°C.

Keywords: emu; pre-pipped embryo; internal pipping; external pipping; cooling; instantaneous heart rate; endothermic response; thermo-conformity response

Introduction

In the chicken (*Gallus gallus domesticus*), late pre-pipped and externally pipped embryos show only a small thermoregulatory metabolic response to cooling (Freeman, 1964, 1971, Tazawa *et al.*, 1988, 1989a, 1989b, Black and Burggren, 2004). This endothermic thermoregulatory competence improves during hatching, and after hatching a drop of hatchling's body temperature during cooling is considerably mitigated by an increase in metabolic rate (Freeman, 1967; Wekstein and Zolman, 1969; Tazawa and Rahn, 1987; Tazawa *et al.*, 2004).

*Corresponding author: TAZAWA, Hiroshi
E-mail: tazawa@mmm.muroran-it.ac.jp

As a result, the initial, poor endothermic competence in late pre-pipped and pipped chicken embryos improves markedly during hatching to provide hatchlings with the ability to maintain endothermy. However, the complete endothermy is not obtained at hatching in the chicken and endothermic competence still improves during an early period of post-hatch life (Tazawa and Rahn, 1987; Tazawa *et al.*, 2004).

Meanwhile, there is evidence that instantaneous heart rate (IHR) and HR baseline in late embryos and hatchlings of chicken and emu respond to cooling and warming in an endothermic fashion (Tamura *et al.*, 2003; Tazawa *et al.*, 2001, 2002). The developmental timeline of endothermic response of IHR has been investigated in hatchlings of broiler chickens (Khandoker *et al.*, 2004; Yoneta *et al.*, 2006; accepted in WPSJ). The investigation showed that the HR response to an alteration of ambient temperature (Ta) changed from a temperature dependent response (i.e., thermo-conformity response) to an inverse temperature dependent response (i.e., endothermic response) during the first day of hatching (Day 0). This suggests that endothermic competence continues to develop on Day 0 in the chicken.

In contrast, emu hatchlings already possess endothermic response of HR upon hatching (Tamura *et al.*, 2003). It is assumed that HR responses to altered Ta (ΔTa) in the emu depend on the state of maturity of thermoregulatory competence as in the chicken. In contrast to the chicken, it appears that the ability for an endothermic response of HR already exists prior to hatching in emu. In the present experiment, we measured HR responses to ΔTa of 10°C in pre-pipped and pipped embryos to elucidate the timing for the development of thermoregulatory HR response in emu.

Materials and methods

EMBRYOS

Fresh fertile eggs were collected in breeding yards of the Cross Timbers Emu Ranch, Flower Mound and the Pearson Emu Ranch, Pilot Point, Texas. The eggs were transported to the Muroran Institute of Technology, Japan, weighed and incubated at a temperature of 36.5°C and relative humidity of about 40% in a forced draught incubator for the present experiment and for other studies. The first 24-hour period of incubation was designated as Day 0. Egg turning was manually made three times a day until experiments were started on Day 46 or Day 47 of incubation. Most embryos tended to internally pip the chorioallantoic and inner shell membranes (internal pipping, referred to as IP) on Days 48-49 of incubation with a few exceptions on Day 47. IP was identified by auscultation of embryo vocalization. Subsequently to IP, embryos externally pipped the eggshell (external pipping, EP) on about Days 49-50 and hatched about half a day later. Experiments were made with embryos prior to IP (referred to as pre-IP embryos) on Days 46-47, those during IP (IP embryos) on Days 47-49 and EP (EP embryos) on Days 49-50 of incubation.

MEASUREMENTS OF EGG TEMPERATURE AND HEART RATE

Eggs were drilled with four holes 3-mm in diameter into the eggshell, taking care not to injure the embryos. Three holes formed a triangle with a side of about 5 cm on the top of the egg for measurement of electrocardiogram (ECG) and the remaining one was made at an arbitrary position close to the triangle for measurement of egg temperature (Te). A thermistor probe 1 mm in diameter was inserted into the egg about 5 mm deep and the thermistor wire in the hole was fixed to the eggshell with clay and epoxy glue. ECG electrodes were silver wires, 5 cm long and 1 mm in diameter, and one end of the wire about 1 cm long was bent so that it could be inserted into the egg under the eggshell. After insertion, three electrodes were fixed to the eggshell with clay and epoxy glue. The eggs treated with the thermistor probe and ECG electrodes were put in a forced-draft, temperature-controlled measuring incubator of which temperature was set first at 36°C. Another thermistor probe was fixed in the measuring incubator to measure Ta.

Basically, IHR of emu embryos was determined from amplified and band-pass filtered ECG signal using a personal computer as described previously for chicken embryos and hatchlings (Moriya *et al.*, 1999, 2000). The ECG electrodes were connected to an amplifier and band-pass filter. The ECG signal was digitized at sampling frequency of 4000 Hz and displayed on a computer screen. The cut-off frequency of the band-pass filter was provided with a sharp slope of 24 octave/dB and programmable range which could be changed by 0.1Hz. Empirically, in chicken embryos the amplitude of ECG waves was very variable and the QRS signals were frequently smaller than T waves or T-like signals which were eliminated by a band-pass filter. Accordingly, in the present emu embryos the frequency range of filter and the amplification of the amplifier were also determined frequently by monitoring R deflections on the computer screen. The band-pass filter with a cut-off frequency between about 30 and 300 Hz generally augmented R deflections above other waves and background noises. Then, a threshold level was set above background noises to detect the raising deflection of R waves on the screen. The time intervals between adjacent R waves which exceeded first the threshold level were converted to IHR in beats per min (bpm) and stored successively in a data file of the computer. On the computer screen, IHR was plotted every heartbeats and the amplified and band-pass filtered R waves were displayed with threshold line. The amplification of amplifier and the cut-off frequency of the band-pass filter were often reset so that the threshold line would not fail to cross the R waves.

For comparison of HR responses, IHR should be averaged to yield mean values of baseline HR. It is likely that baseline HR is presented by mode heart rate (MHR) better than mean HR, because mean values are apt to be affected by transient HR decelerations from baseline. Therefore, MHR was used for statistical analysis and comparison of HR responses in order to exclude possible effect of transient deviations of IHR on mean HR. However, mean HR was also calculated to show any possible adverse effect of transient HR outliers on comparison of HR responses.

PROCEDURE OF EXPOSURE

The eggs were kept in the measuring incubator for at least 1.5 hours prior to the measurement for temperature equilibration at Ta of 36°C. Both Te and IHR were continuously recorded on the computer file. After the change in Te became less than 0.1°C during a 30-min period, measurement at Ta of 36°C was started and continued for the next 90 minutes. Then, Ta of the measuring incubator was changed to 26°C for another 90-min period with subsequent recovery to 36°C for a further 90 minutes. The measurement was made for 270-min period in total. As a result, the eggs were consecutively exposed to Ta of 36°C, 26°C and 36°C for 90-min period each; i.e., sequence of cooling and warming.

STATISTICAL ANALYSIS

One-way factorial ANOVA examined the differences in Te (or HR) between the three developmental stages; i.e., pre-IP, IP and EP, at an individual Ta; i.e., 36°C, 26°C and 36°C, using a fiducial level of $P<0.05$. When ANOVA revealed significant differences, Tukey post hoc test was used to detect the differences between group means. For each developmental stage, the differences in Te (or HR) between three exposures to Ta of 36°C, 26°C and 36°C were examined by one-way repeated measures ANOVA for the significance at $P<0.05$. Tukey post hoc test detected the differences of means between the three temperature exposures. Two-way factorial ANOVA was used to examine the differences between mode heart rate (MHR) and mean HR and the differences of HR between the three temperature exposures with Tukey post hoc test. The mean values of Te and MHR were shown with standard error (SE).

Results

CONTINUOUS MEASUREMENTS

Measurements of Te and IHR during exposures to ΔTa of 10°C were made for 12 embryos. Among them, three embryos were repeatedly examined at pre-IP, IP and EP stages and the remaining nine embryos were tested once during either pre-IP, IP or EP period for a total of six animals per developmental stage. The mean mass of 12 eggs prior to incubation was 614 ± 18 g.

Figure 1 shows 270-min recordings of Te and IHR in six pre-IP embryos which were consecutively exposed to Ta of 36°C, 26°C and 36°C. Te and IHR of IP and EP embryos are presented in Figures 2 and 3, respectively. The recordings shown in the top panel of Figures 1, 2 and 3 were measured from the same embryo; that is, the embryo N1. Similarly, those in the second and third panels of Figures 1, 2 and 3 correspond to the embryo N2 and the embryo N3, respectively. The changes in Ta upon switching at 90 min and 180 min were attained within 5 min.

In pre-IP embryos (Figure 1), Te which was higher than Ta during pre-cooling reference period gradually decreased during cooling, but did not reach Ta by the end of the 90 min cooling period. Upon warming, Te increased during the entire 90-min period and never completely recovered the control levels. In conformity with the changes in Te, HR baseline was stable during the pre-cooling period, then continuously decreased during cooling and increased slowly during warming with only partial recovery.

In IP embryos (Figure 2), the changes in Te in response to altered Ta were similar to those observed in the pre-IP embryos. Te gradually decreased during cooling. However, unlike the pre-IP embryos, HR baseline of all IP-embryos except one (N9) transiently increased upon cooling with a subsequent gradual decrease during the remaining cooling period. HR baseline then showed inconsistent responses during warming, increasing in some embryos and remaining relatively constant in others.

In embryos that had externally pipped the eggshell, the responses of Te and HR were different from pre-IP and IP embryos (Figure 3). Mean pre-cooling control Te tended to be higher than that of IP embryos and the decrease in Te during cooling was mitigated in most EP embryos. In embryos N2, N3 and N10, Te did not change more than 0.1°C during the final 30 min of cooling. Additionally, the recovery of Te during warming occurred rapidly.

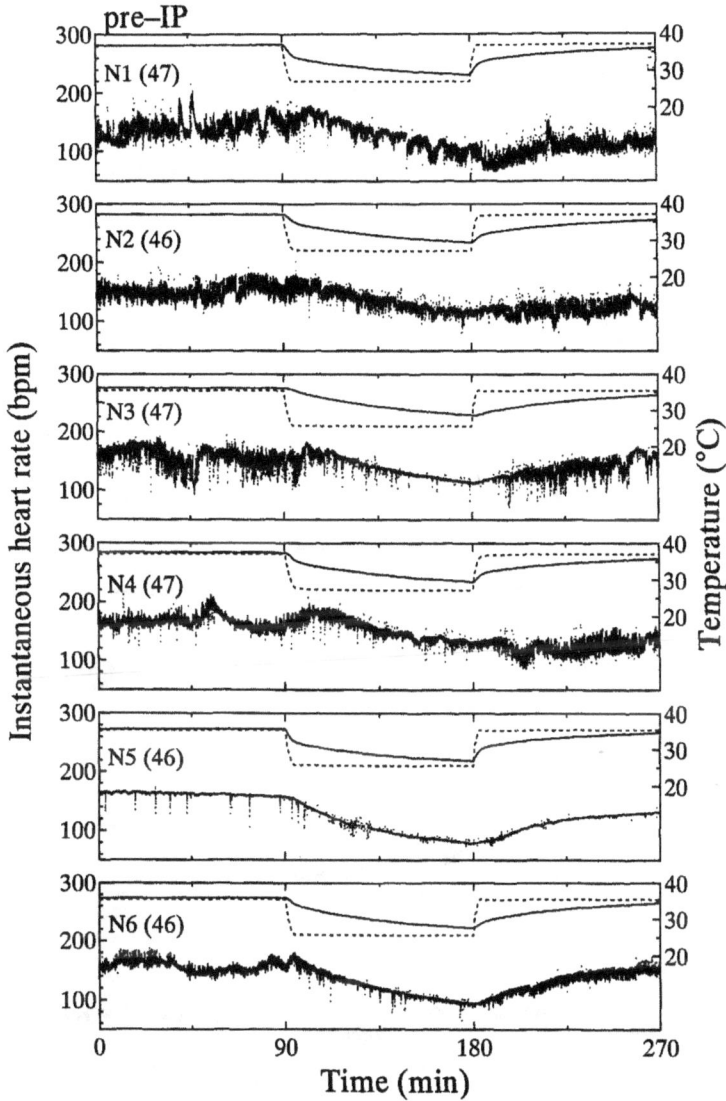

Figure 1. Responses of instantaneous heart rate (shown by individual points) and egg temperature (Te) to altered ambient temperatures (Ta; 36°C, 26°C and 36°C) in six pre-IP embryos which were numbered from N1 (top) to N6 (bottom). Te is shown by solid line and Ta, dotted line. The number in the parentheses indicates the incubation day when the measurement was made. The scale of temperature was shown on the right side. The recording of Te during the first 90-min period (pre-cooling reference period) was flat and exceeded slightly above Ta.

In these embryos, IHR baseline, which initially increased in response to cooling, remained elevated during the cooling exposure. Inverse temperature dependent response of IHR baseline upon cooling also occurred in remaining embryos (N1, N11 and N12). Upon warming, IHR baseline of embryos N1, N2 and N10 swiftly returned to the pre-cooling control level, while that of embryos N3 and N11 remained elevated. In embryo N12, IHR baseline decreased with time during cooling and tended to increase during warming.

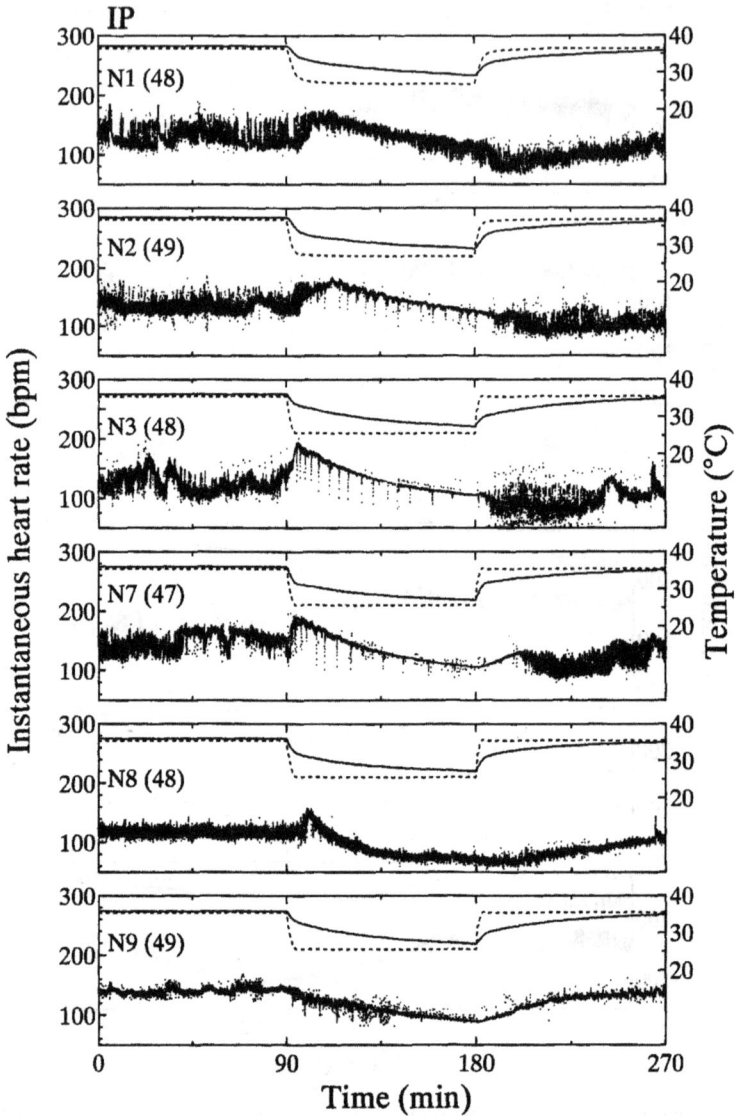

Figure 2. Responses of instantaneous heart rate (points) and egg temperature (solid line) to altered ambient temperatures (dotted line) in six IP embryos. Three embryos from the top; N1, N2 and N3, were the same as those in Figure 1.

RESPONSES OF EGG TEMPERATURE AND MODE HEART RATE

Because Te of most embryos did not reach a steady state during 90-min period of cooling and warming exposures, differences of responses between the three developmental stages were examined by comparison of mean values during the last 10-min exposure to each Ta. Accordingly, the last 10-min values of Te during exposures to 36°C, 26°C and 36°C were averaged in individual embryos and referred to as Te_{ref}, Te_{26} and Te_{36}, respectively. For comparison of HR responses, mode heart rate (MHR) was calculated for IHR recording

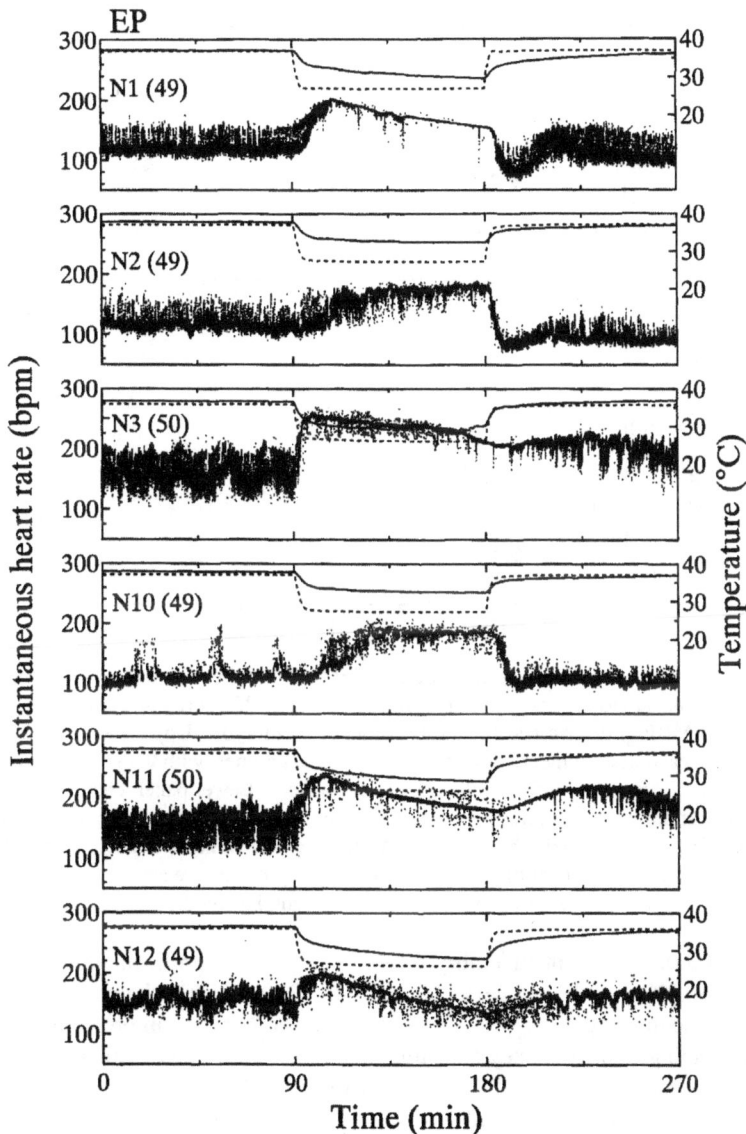

Figure 3. Responses of instantaneous heart rate (points) and egg temperature (solid line) to altered ambient temperatures (dotted line) in six EP embryos. Among them, three embryos; N1, N2 and N3, were the same as those in Figures 1 and 2.

during the last 10-min period of individual exposures (i.e., MHR_{ref}, MHR_{26} and MHR_{36}). Figure 4 shows the response of Te to ΔTa; that is, Te_{ref}, Te_{26} and Te_{36}, and the response of MHR; MHR_{ref}, MHR_{26} and MHR_{36}, in pre-IP, IP and EP embryos.

Te_{ref} was 36.6 ± 0.3, 36.4 ± 0.3 and $37.0\pm0.2°C$ for pre-IP, IP and EP embryos, respectively. Te_{ref} of EP embryos was highest, but the difference was not significant compared with pre-IP and IP embryos (P=0.316). Te_{26} of EP embryos ($29.7\pm0.6°C$) was also high compared with that of pre-IP ($28.3\pm0.4°C$) and IP embryos ($28.0\pm0.4°C$), but the difference was not significant (P=0.099). However, Te_{36} was significantly different between the three

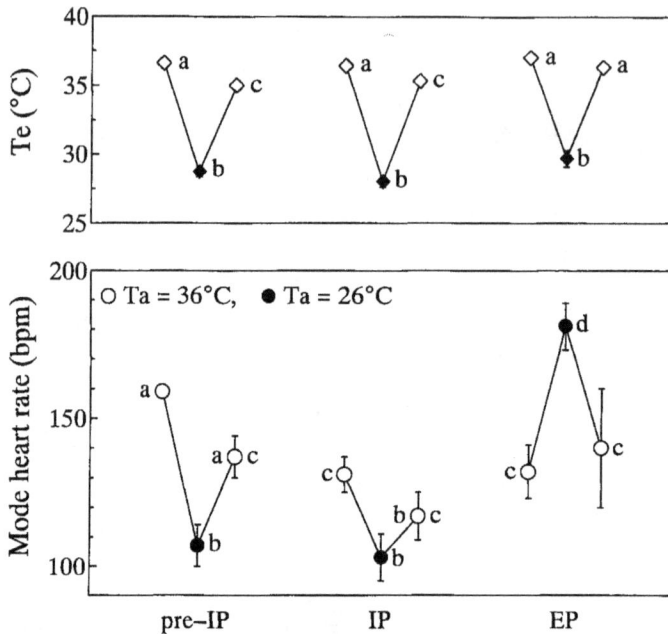

Figure 4. Mean value with SE of egg temperature (Te) and mode heart rate (MHR) during the last 10-min period of exposure to the first 36°C (Te$_{ref}$ and MHR$_{ref}$) and 26°C (Te$_{26}$ and MHR$_{26}$) and the second 36°C (Te$_{36}$ and MHR$_{36}$) determined for pre-IP, IP and EP embryos. Upper panel: mean Te with SE (vertical bar) in each developmental group connected by solid lines, with ◊ on the left indicating Te$_{ref}$, ◆ in the center showing Te$_{26}$ and ◊ on the right showing Te$_{36}$. Lower panel: changes in MHR during cooling and warming with solid lines connecting mean value in each developmental group. Vertical bar indicates SE. ○ on the left show MHR$_{ref}$, ● in the center indicate MHR$_{26}$, and ○ on the right indicate MHR$_{36}$. The significant differences between group means are shown by different symbols, and if the differences are not significant, those groups are marked by the same symbol.

stages (P=0.009); i.e., Te$_{36}$ of EP embryos (36.3±0.3°C) was significantly higher than that of pre-IP (35.0±0.3°C) and IP (35.3±0.2°C) embryos (P<0.05 for both comparisons). Within each developmental stage, Te$_{36}$ was significantly lower than Te$_{ref}$ in pre-IP and IP embryos (P<0.01 for both groups by Tukey test, after significant difference, P<0.0001, by ANOVA), but in EP embryos Te$_{36}$ was not different from Te$_{ref}$ (P>0.05).

MHR$_{ref}$ in pre-IP, IP and EP embryos was 159±2, 131±6 and 132±9 bpm, respectively, which were significantly different (P=0.014). MHR$_{ref}$ of pre-IP embryos was significantly higher than that of IP and EP embryos (P<0.05 for both comparisons). While cooling decreased MHR$_{26}$ of pre-IP and IP embryos (107±7 and 103±8 bpm, respectively), EP embryos increased markedly MHR$_{26}$ (181±8 bpm) which was much higher than the former two (P<0.01 for both comparisons). Meanwhile, MHR$_{36}$ was not different between three developmental stages; i.e., 137±7, 117±8 and 140±20 bpm for pre-IP, IP and EP embryos, respectively (P=0.434).

In pre-IP embryos, MHR$_{ref}$, MHR$_{26}$ and MHR$_{36}$ were significantly different (P=0.0005). The difference between MHR$_{ref}$ and MHR$_{26}$ and that between MHR$_{26}$ and MHR$_{36}$ were significant (P<0.01 and P<0.05, respectively), but the difference between MHR$_{ref}$ and MHR$_{36}$ was not significant. In IP embryos, MHR was significantly different between three temperature environments (P=0.027); i.e., the difference between MHR_{ref} and MHR_{26} was significant (P<0.05), but the difference between MHR_{26} and MHR_{36} and that between MHR_{ref} and MHR_{36} were not significant. In EP embryos, MHR was significantly different between three temperature environments (P=0.009); i.e., MHR_{26} was significantly higher than MHR_{ref} (P<0.05) and MHR_{36} (P<0.05), but the latter two were not different each other.

Mean HR was calculated for the same last 10-min recording of IHR as used for calculation of MHR in individual embryos shown in Figures 1, 2 and 3. Table 1 summarizes mean values of mean HR with SE in six embryos at pre-IP, IP and EP stages and also those of MHR for comparison. The latter data are the same as in Figure 4. The statistical analysis revealed that mean HR was not significantly different from MHR.

Table 1. Summary of mean heart rate and mode heart rate at altered Ta in pre-IP, IP and EP emu embryos (N=6).

Stage		Pre-IP			IP			EP		
Ta		36°C	26°C	36°C	36°C	26°C	36°C	36°C	26°C	36°C
Mean HR	Mean	156[a,c]	105[b]	135[c]	133[a,c]	103[b]	120[b,c]	138[a,c]	179[b]	141[c]
	SE	±2	±7	±7	±5	±8	±6	±9	±9	±19
Mode HR	Mean	159	107	137	131	103	117	132	181	140
	SE	±2	±7	±8	±6	±8	±8	±9	±8	±20

At individual stages, the same symbol means that the group means are not significantly different and different symbols indicate significant difference.

Discussion

CRITIQUES OF EXPERIMENTAL PROCEDURES

When avian eggs are exposed to cooling, they take some time to equilibrate to a new steady state temperature (quasi-equilibrium temperature); e.g., about 3 hours in chicken eggs and 12 hours in ostrich eggs (Tazawa *et al.*, 2001). Because the mass of emu eggs falls between these two species, it is predicted that cooling to a quasi-equilibrium temperature will take some time in-between. Thus a 90 min exposure is too short for emu eggs to reach the quasi-equilibrium temperature. In most pre-IP and IP embryos, Te was still decreasing during the last 10-min period of cooling and did not return to the level of initial reference Te during warming (Figs. 1, 2 and 4). Three EP embryos (N2, N3 and N10) approached the quasi-equilibrium Te during the 90-min cooling period. In three other EP embryos (N1, N11 and N12), Te continuously decreased during cooling. This resulted in the mean value of Te_{26} in six EP embryos similar to that in pre-IP and IP embryos. However, in these EP embryos the mean value of Te_{36} returned to that of Te_{ref}, which was not the case in pre-IP and IP embryos (Fig. 4).

A glimpse of the difference in response of Te between EP embryos and pre-IP and IP embryos was obtained by 90-min exposing procedures before reaching quasi-equilibrium Te. Because embryos continue to develop from pre-IP stage to IP and EP stages, prolonged exposure lasting several hours may occasionally extend over two different stages which produce different response of Te (and HR) to ΔTa. Contrarily, a shortened period of exposure will produce only a transient response, which may fail to present correctly development of thermoregulatory competence. Because it is not proved whether a 90-min exposure is adequate, development of endothermic HR response was evaluated on the condition that the experiment was made on the basis of 90-min exposures.

HR responses to cooling were determined by exposing embryos to Ta lowered by 10°C (ΔTa=-10°C). If embryos were exposed to another low Ta, for instance, ΔTa=-5°C or -20°C, HR responses might be different from the present result. An increase of HR baseline in IP embryos exposed to ΔTa=-5°C may be more pronounced than the present transient increase induced by ΔTa=-10°C, and the present pronounced increase of HR baseline in EP embryos exposed to ΔTa=-10°C may not be obtained by exposure to ΔTa=-20°C. Accordingly, development of endothermic HR response was evaluated on the condition that the experiment was made on the basis of exposure to ΔTa=-10°C and 10°C.

The exposure sequence in the present experiment was cooling and warming periods. In chickens, it was proved that evaluation of development of endothermic HR response was not affected by exposure sequence (Yoneta *et al.*, 2006, accepted in WPSJ). Therefore, it is thought that the opposite sequence of warming and cooling exposure yields the same developmental timeline of endothermic HR response in emu embryos.

DEVELOPMENT OF ENDOTHERMIC HEART RATE RESPONSE

In addition to the measurements that examined once the HR responses to cooling and warming exposure in nine embryos, three embryos were subjected to repeated exposures to cooling and warming during the pre-IP, IP and EP periods. Although quantitative comparisons between single measurements and repeated measurements could not be made because of lack of enough data, repeated exposures to cooling and warming are not likely to affect development of endothermic HR response. Because changes in HR responses with development were visually similar between repeated group (N1, N2 and N3) and non-repeated group (N4-N12), both the data were pooled to evaluate the development of endothermic HR response.

Exposure of pre-IP embryos to ΔTa= -10°C and then 10°C induced the thermo-conformity response of HR. IHR baseline and Te began to decrease with exposure to cooling and tended to restore pre-cooling reference level during warming (Figs. 1 and 4). Responses in IP embryos became different from pre-IP embryos with regard to the early transient increase in HR baseline upon cooling and insignificant increase in HR during warming (Figs. 2 and 4). These responses in IP embryos suggest that some development of thermoregulatory competence occurs during IP. In EP embryos, the difference of responses from pre-IP and IP embryos became much distinctive, particularly in HR responses. Although Te_{26} of EP embryos was not significantly kept high compared with that of pre-IP and IP embryos, Te_{36} returned to Te_{ref} in EP embryos (Fig. 4). This response of Te in EP embryos, which was slightly different from pre-IP and IP embryos, seems to suggest that development of thermoregulatory competence progresses during EP. Contrarily, with regard to HR, responses to cooling depicted the inverse temperature dependent (i.e., endothermic) pattern (Figs. 3 and 4), which were distinctively different from the temperature dependent (thermo-conformity) pattern of pre-IP and IP embryos. While the development of thermoregulatory competence is slightly exhibited by changes of Te in EP embryos, distinctive turnover of HR response from thermo-conformity to endothermic pattern must be attributed to a change in thermoregulatory competence during EP period in the emu.

The quick response of IHR may be at least partially neurogenic, but it remains to be studied how autonomic nervous functions are related to regulate the HR responses. Crossley *et al.* (2003) showed that the adrenergic and cholinergic tones of emu embryos developed as early as 70% of incubation and continued to increase through 90% of incubation. Additionally, the day old emu hatchling exhibits the same levels of adrenergic and cholinergic tones on HR (Dzialowski, unpublished data). Thus, it appears that the autonomic nervous control of HR is in place prior to IP, but does not alter HR during cooling and warming by ΔTa of 10°C until EP. A lack of endothermic pattern in two EP embryos, N3 and N11, during warming may be attributed to a less developed thermoregulatory competence in these embryos or suggests an unknown contribution to HR responses.

In chickens, 18-day-old pre-IP embryos decreased exponentially Te when exposed to ΔTa of about -10°C from 38°C (Tazawa and Nakagawa, 1985) and EP embryos also decreased Te and HR in an exponential fashion (Tazawa and Rahn, 1986, 1987). Judging from these previous experiments, HR responses to ΔTa=-10°C in chickens are considered to exhibit thermo-conformity pattern during embryonic development, even in EP embryos. On the day of hatching, chickens exhibited the turnover from thermo-conformity to endothermic response of HR (Khandoker *et al.*, 2004; Yoneta *et al.*, 2006, accepted in WPSJ). It is likely that development of thermoregulatory competence is more advanced in the emu than in the chicken.

We assumed that transient accelerations and decelerations of IHR might affect mean values of baseline HR to a higher degree than mode values and thus used MHR to evaluate development of endothermic HR response in the present study. At the same time, we calculated mean HR to show the effect of transient IHR deviations on baseline HR. However, group means of mean HR were not significantly different from those of MHR, indicating that both mean values and mode values can be used to present baseline HR in emu embryos.

INSTANTANEOUS HEART RATE

In late emu embryos developing at the incubation temperature of 36.5°C, IHR fluctuations were substantially composed of irregular accelerations and three main patterns of IHR accelerations were previously reported (Kato *et al.*, 2002). They were irregular intermittent large IHR accelerations, wide accelerations around the baseline, and short-term repeated large accelerations. These IHR acceleration patterns were also recorded in the present measurement at 36°C, but exposure to cooling eliminated these fluctuations to produce a narrow HR baseline in many embryos. Figure 5 shows an example taken from EP embryo, N1, presented in Figure 3. The irregular intermittent large accelerations during the last 10-min pre-cooling period turned to frequently repeated accelerations producing a raise of the baseline during the first 10-min period of cooling and afterward the accelerations were eliminated as the baseline increased during the continued cooling (top panel). The beat-to-beat changes in IHR were almost eliminated during the last 10 min of cooling and fluctuations began to appear again with warming which decreased the HR baseline (bottom panel). Transient HR decelerations remained during cooling in this embryo and also several other embryos at each developmental stage. Because HR decelerations are mediated by vagus nerve in chicken embryos (Höchel *et al.*, 1998; Chiba *et al.*, 2004), it seems that the vagus nerve is functioning during cooling in the emu. At this point in development, the emu embryo has been shown to have a functioning vagal tone (Crossley *et al.*, 2003). In EP embryos, elimination

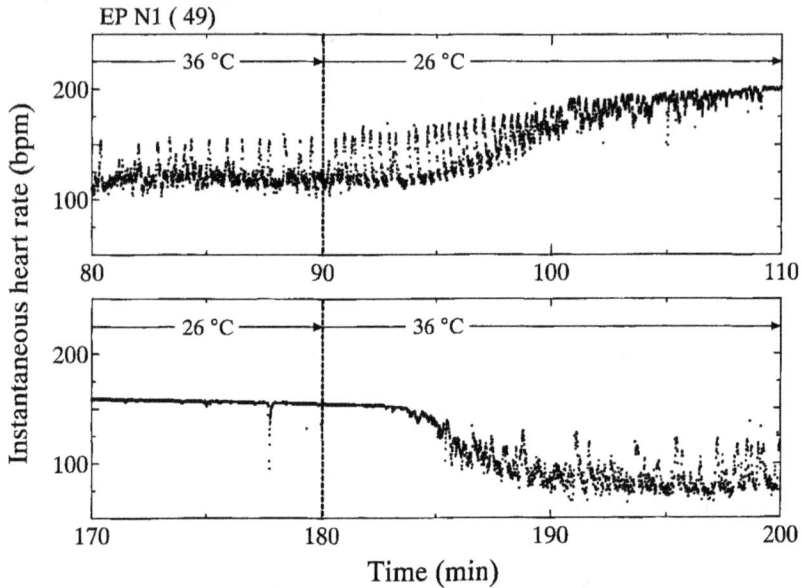

Figure 5. Thirty-min recordings of instantaneous heart rate in an externally pipped embryo which was exposed to altered ambient temperature from 36°C to 26°C (top panel) and from 26°C to 36°C (bottom panel). These time-expanded recordings are taken from the top panel of Figure 3. The abscissa indicates the time (in min) elapsing from the beginning of recording and vertical dotted lines show the time of change in Ta.

of IHR accelerations during cooling seems to be caused by decapitation of high HR. However, IHR accelerations were eliminated in pre-IP and IP embryos of which HR baseline was not raised by cooling.

Emu embryos develop endothermic HR response prior to hatching. This endothermic response develops later in the chicken throughout the first day of hatching (Khandoker *et al.*, 2004; Yoneta *et al.*, 2006, accepted in WPSJ). Mechanisms of the shift from thermoconformity to endothermic response of HR in both the chicken and emu remain to be studied.

Acknowledgments

We are grateful to Ms. Pearson for supplying the eggs from the two emu ranches at Flower Mound and Pilot Point in Texas. This study was supported in part by National Science Foundation Grant No.IOB0417205 awarded to EMD and No.IBN0128043 to WWB and by a Grant-in-Aid for Scientific Research of the Japan Society for Promotion of Science (No. 15560352) to HT for collaboration with EMD and WWB.

References

BLACK, J.L. and BURGGREN, W.W. (2004) Acclimation to hypothermic incubation in developing chicken embryos (*Gallus domesticus*) I. Developmental effects and chronic and acute metabolic adjustments. *Journal of Experimental Biology* **207**: 1543-1552.

CHIBA, Y., FUKUOKA, S., NIIYA, A., AKIYAMA, R. and TAZAWA, H. (2004) Development of cholinergic chronotropic control in chick (*Gallus gallus domesticus*) embryos. *Comparative Biochemistry and Physiology* **A 137**: 65-73.

CROSSLEY, D.A., BAGATTO, B.P., DZIALOWSKI, E.M. and BURGGREN, W.W. (2003) Maturation of cardiovascular control mechanisms in the embryonic emu (*Dromiceius novaehollandiae*). *Journal of Experimental Biology* **206**: 2703-2710.

FREEMAN, B.M. (1964) The emergence of the homeothermic-metabolic response in the fowl (*Gallus domesticus*). *Comparative Biochemistry and Physiology* **13**: 413-422.

FREEMAN, B.M. (1967) Some effects of cold on the metabolism of the fowl during the perinatal period. *Comparative Biochemistry and Physiology* **20**: 179-193.

FREEMAN, B.M. (1971) Body temperature and thermoregulation. In: *Physiology and Biochemistry of the Domestic Fowl.* (Ed. Bell, D.J. and Freeman, B.M.), Academic Press, New York, pp.1115-1151.

HÖCHEL, J., AKIYAMA, R., MASUKO, T., PEARSON, J.T., NICHELMANN, M. and TAZAWA, H. (1998) Development of heart rate irregularities in chick embryos. *American Journal of Physiology* 275 (*Heart Circ. Physiol.*) **44**: H527-H533.

KATO, K., MORIYA, K., DZIALOWSKI, E., BURGGREN, W.W. and TAZAWA, H. (2002) Cardiac rhythms in prenatal and perinatal emu embryos. *Comparative Biochemistry and Physiology* **A 131**: 775-785.

KHANDOKER, A.H., FUKAZAWA, K., DZIALOWSKI, E.M., BURGGREN, W.W., and TAZAWA, H. (2004) Maturation of the homeothermic response of heart rate to altered ambient temperature in developing chick hatchlings (*Gallus gallus domesticus*). *American Journal of Physiology* (*Regul. Integr. Comp. Physiol.*) **286**: R129-R137.

MORIYA, K., HÖCHEL, J., PEARSON, J.T. and TAZAWA, H. (1999) Cardiac rhythms in developing chicks. *Comparative Biochemistry and Physiology* **A 124**: 461-468.

MORIYA, K., PEARSON, J.T., BURGGREN, W.W., AR, A. and TAZAWA, H. (2000) Continuous measurements of instantaneous heart rate and its fluctuations before and after hatching in chickens. *Journal of Experimental Biology* **203**: 895-903.

TAMURA, A., AKIYAMA, R., CHIBA, Y., MORIYA, K., DZIALOWSKI, E.M., BURGGREN, W.W. and TAZAWA, H. (2003) Heart rate responses to cooling in emu hatchlings. *Comparative Biochemistry and Physiology* **A 134**: 829-838.

TAZAWA, H. and NAKAGAWA, S. (1985) Response of egg temperature, heart rate and blood pressure in the chick embryo to hypothermal stress. *Journal of Comparative Physiology* **B 155**: 195-200.

TAZAWA, H. and RAHN, H. (1986) Tolerance of chick embryos to low temperatures in reference to the heart rate. *Comparative Biochemistry and Physiology* **85A**: 531-534.

TAZAWA, H. and RAHN, H. (1987) Temperature and metabolism of chick embryos and hatchlings after prolonged cooling. *Journal of Experimental Zoology* **Suppl. 1**: 105-109.

TAZAWA, H., WAKAYAMA, H., TURNER, J.S. and PAGANELLI, C.V. (1988) Metabolic compensation for gradual cooling in developing chick embryos. *Comparative Biochemistry and Physiology* **89A**: 125-129.

TAZAWA, H., OKUDA, A., NAKAZAWA, S. and WHITTOW, G.C. (1989a) Metabolic responses of chicken embryos to graded, prolonged alterations in ambient temperature. *Comparative Biochemistry and Physiology* **92A**: 613-617.

TAZAWA, H., WHITTOW, G.C., TURNER, J.S. and PAGANELLI, C.V. (1989b) Metabolic responses to gradual cooling in chicken eggs treated with thiourea and oxygen.

Comparative Biochemistry and Physiology **92A**: 619-622.

TAZAWA, H., MORIYA, K., TAMURA, A., KOMORO, T. and AKIYAMA, R. (2001) Ontogenetic study of thermoregulation in birds. *Journal of Thermal Biology* **26**: 281-286.

TAZAWA, H., MORIYA, K., TAMURA, A. and AKIYAMA, R. (2002) Low frequency oscillation of instantaneous heart rate in newly hatched chicks. *Comparative Biochemistry and Physiology* A **131**: 797-803.

TAZAWA, H., CHIBA, Y., KHANDOKER, A.H., DZIALOWSKI, E.M. and BURGGREN, W.W. (2004) Early development of thermoregulatory competence in chicken; responses of heart rate and oxygen uptake to altered ambient temperatures. *Avian and Poultry Biology Reviews* **15**: 166-176.

WEKSTEIN, D.R. and ZOLMAN, J.F. (1969) Ontogeny of heat production in chicks. *Federation Proceedings* **28**: 1023-1028.

YONETA, H., FUKAZAWA, K., DZIALOWSKI, E.M., BURGGREN, W.W. and TAZAWA, H. (2006) Does sequence of exposure to altered ambient temperature affect the endothermic heart rate response of newly hatched chicks? In: *New Insights into Fundamental Physiology and Peri-natal Adaptation of Domestic Fowl*. Nottingham University Press, Nottingham, pp15-28.

Effect of a reduced environmental oxygen content on the development of the myocardium in chick embryos

SASCIA LANGE[1]*, RUTH HIRSCHBERG[1], HEIKE TÖNHARDT[2] AND HERMANN BRAGULLA[3]

[1] Institute of Veterinary Anatomy, Faculty of Veterinary Medicine, Freie Universität Berlin, Koserstraße 20, D-14195 Berlin, Germany; [2] Institute of Veterinary Physiology, Faculty of Veterinary Medicine, Freie Universität Berlin, Oertzenweg 19 b, 14163 Berlin, Germany; [3] Department of Biological Sciences, Louisiana State University, 202 Life Science Building, Baton Rouge, La 70803-1715, USA

While developing from a mesoderm-derived tube-like organ to a complex four-chambered organ, the embryonic heart – complying with the growing embryo's needs – develops an increased contractive force as well as alterations in function and effectiveness. A like-wise 'increased efficiency' of the heart is potentially needed under a reduced environmental oxygen content that requires qualitative alterations of the cardiac cells, such as differences in the degree and time scale of differentiation by enhancing the synthesis of contractile elements of the cytoskeleton, i.e. hypertrophy. It might also induce quantitative alterations within the contracting syncyticum of cardiomyocytes, i.e. hyperplasia. A multitude of studies support the view that the chick embryo – developing within the egg – is endowed with a high tolerance for oxygen deficiency conditions. This tolerance results from the various adaptive abilities of the embryonic organism. Complementing the physiologic studies of Decker (2002) and DzialOwski (2002), the present morphologic study renders new insights into the effect of the incubation under reduced environmental oxygen content on the development of the embryonic heart, particularly regarding its cellular and structural components.

The morphometric examination of this study showed that quantitative alterations prevailed within the adaptive mechanisms of the heart. The chick embryos incubated between day 6 and day 12 under reduced environmental oxygen content, i.e. 15 % O_2 instead of 21 % O_2, displayed an increase of the mean value of the cell numbers regarding three different locations in the heart (left ventricle, interventricular septum, and right ventricle) compared to the respective locations within the control group. Although these results were only statistically significant for the interventricular septum, they indicate that the reduced environmental oxygen content indeed affects the structural development and differentiation of the embryonic myocardium. This adaptive mechanism results presumably from a "reactive" increased mitotic activity in the myocardium, i.e. hyperplasia, after a "sensitive phase" of the chick embryo between the sixth and twelfth day of incubation, respectively.

Keywords: Chick heart, myocardium, morphometry, hyperplasia, hypertrophy, embryo

*Corresponding author
E-mail: Sascia897@compuserve.de

This study is part of the dissertation thesis of Sascia Lange (Effect of a reduced oxygen content on the development of the myocardium in chick embryos. Berlin, Freie Universität, Faculty of Veterinary Medicine, dissertation thesis, 2005)

Introduction

During embryonic development of the chick, myocardial functions are adapted continuously, concomitant with a complex remodelling of the heart. As in other metabolic highly active organs, we expected the developing heart to be sensitive to a reduction of the environmental oxygen content. From previous studies it is known that – depending on the developmental stage, the duration and the degree of oxygen deficiency – the embryonic vascular system and metabolism have the ability to compensate and to prevent an actual hypoxic situation in the embryo. Compared to a diminished environmental oxygen content, those "effective" hypoxic situations cause a deficient supply of the myocardium with oxygen. An oxygen deficiency, affecting specific tissues, may be the consequence of local or systemic alterations in the oxygen supply to the embryonic tissues. Local deficiencies in the oxygen supply may have drastic effects on the myocardial function and development that may lead to heart failure. Compared to those dramatic local effects, systemic alterations in oxygen supply display, due to different compensatory mechanisms, rather long term effects with less impact on cardiac function. Little is known about the time course and effectiveness of long term hypoxia-compensating mechanisms. As in humans and in other mammalian species, many avian species are able to exist under reduced oxygen conditions. Such an adaption to adverse conditions requires protective mechanisms for the organism and necessary adjustments. These include mechanisms that have an influence on the foetal metabolism as well as on the development and those who optimise the foetal oxygen transport.

A number of studies (Dusseau *et al.*, 1989; Dzialowski *et al.*, 2002; Decker, 2002; Hühnke, 2003) proved a high adaptional capacity of chick embryos to external factors like diminished environmental oxygen content. These studies indicate a "critical" or "sensitive phase" of various embryonic tissues during the prenatal development between the sixth and the twelfth day of incubation (Dusseau *et al.*, 1989; Dzialowski *et al.*, 2002).

Complementing previous physiological studies (Decker, 2002), the present morphologic study renders new insights into the effect of a diminished environmental oxygen content on the development of the embryonic chick heart during incubation, particularly regarding its cellular and structural components.

The heart's ability to contract is critically dependent upon a syncytic myocardium built up from a determined number of cardiomyocytes that contain a precise assembly and alignment of myofilaments, i.e. actin and myosin, and other proteins, within the sarcomeres. As oxygen is noted to be a potential regulator for growth and metabolism (Asson-Batres *et al.*, 1989), we assumed structural alterations in the embryonic heart based on the mean values of myocardiac cell numbers, i.e. hyperplasia, in order to provide the higher cardiac efficiency required for the heart of a growing chick embryo under hypoxic conditions.

Material and methods

Fertilized White Leghorn chicken eggs were incubated under normal conditions (21 % O_2, room temperature, moisture 60 % ± 2) in an incubator with an automatic device to turn the eggs.

On day six of incubation, 40 eggs were randomly assigned to each of two groups (control and experimental group). Between day 6 and day 12 of incubation, the 21 eggs of the experimental group were provided with a low environmental oxygen content of 15 % O_2, while the remaining 19 eggs of the control group were incubated in a normal oxygen environment throughout. After a certain time to develop under these defined conditions, i.e. reduced or normal oxygen supply, the eggs were opened after an incubation period of 10, 12, 14, 16, 18 or 20 days, respectively, to take out the embryos. Subsequently the hearts of these different developmental stages were dissected from the embryos using magnifying lenses. Thus extracted hearts for the histologic and morphometric study were generously donated by Decker (2002) from her physiologic research project. A total of 33 embryonic hearts were used for the morphometric study.

Using a microtome, a series of transverse histological sections with a thickness of 5 μm of all the 33 embryonic chick hearts developed under either normoxic or hypoxic conditions (see chart 1) were produced, and those serial sections were examined by light microscopy and used for morphometry. The sections were characterised by routine histologic staining techniques (Trichrome, Periodic-Acid-Schiff-reaction, or Hematoxylin & Eosin) according to Romeis (1989). The numbers of cells in the myocardium were evaluated microscopically by counting the nuclei in a field of vision (rectangle: 90 μm x 70 μm, overall magnification 1000x) in three different locations of the heart, by counting all nuclei (myocardial and non-myocardial cells) for each of the three locations at the level of the atrioventricular valves (see fig. 1) in three consecutive sections taken from the middle of the series of sections to determine the mean values in the numbers of cells in the compact external and mean part of the myocardium (see fig. 2) for the outer wall of the left ventricle, for the interventricular septum, and for the outer wall of the right ventricle. For the graphical and statistical presentations of the results we utilized the statistic program SPSS 12.0 (*SPSS Software GmbH, München, Germany.*)

After collecting the histological and morphometrical data from all samples (blind study), the allocation to the control or experimental groups, respectively, was revealed, and then the data were compared and analysed regarding differences between both groups. For statistical analyses of the morphometric results, the data were tested using the Shapiro-Wilk-test concerning the normal distribution of the cell numbers and using the method of analysis of variance via confidence interval according to Scheffé to verify the significance of the differences in the numbers of myocardial cells between the control group and the experimental group.

Results

The growing chick embryo requires a fast developing embryonic heart (fig. 1) which displays an increasing contractive force as well as alterations in function and effectiveness.

A similar "increased efficiency" of the heart is potentially needed under a reduced environmental oxygen content and thus requires qualitative, i.e. intracellular, alterations of the cardiomyocytes, such as differences in the degree of differentiation and myocardialization, i.e. the tissue-specific characteristics of myocardium. On the other hand, a reduced environmental oxygen content might also induce quantitative alterations within the contracting syncytium of cardiomyocytes.

The morphometric examination of this study showed that quantitative alterations prevailed within the adaptive mechanisms of the embryonic heart (see Tables 1 and 2). Additionally, age related differences in the structure of the embryonic myocardium (see fig. 2 a and b) were detected.

Fig. 1. Overwiew of an embryonic chick heart after 14 days of incubation (control group). The plotted squares exemplify the sample sites chosen for the microscopic examination within the three different locations (A, B, and C); transverse section, left ventricle to the right, right ventricle to the left.

Legend: A = left ventricle myocardium, B = interventricular septum, C = right ventricle myocardium, D = left atrium, E = right atrium, original magnification: 50x

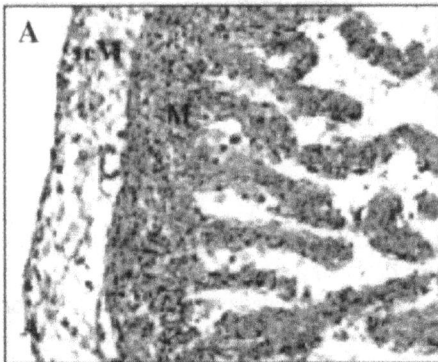

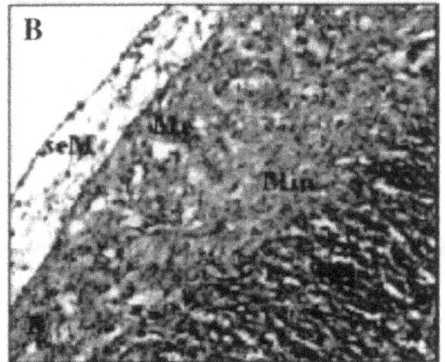

Fig. 2. The development of the cardiac tissue in the ventricular wall: Comparison between the structures of the embryonic chick myocardium at different stages of development under normal conditions, i.e. control group, to demonstrate the structural changes correlated to the time period of development. After nine days of incubation (fig. 2 A), a trabecular, undirectional structure of the left ventricle myocardium is detectable. At the same location, a non-trabecular, but well organised and multilayered arrangement of cardiomyocytes was detected after 16 days of incubation (fig. 2 B).

Legend: seM = subepicardiac mesenchyme, M = trabecular myocardium, Me = external myocardial layer, Mm = mean myocardial layer, Mi = internal myocardial layer; Original magnification: 200x

After a period of 10 or 12 days, respectively, the mean value of the numbers of cells in the myocardium of the outer wall of the left ventricle, of the interventricular septum, and of the outer wall of the right ventricle in both the control and the experimental groups are almost in the same range. Comparing the mean values of cell numbers in the three distinct locations of the embryonic myocardium in both groups after the incubation period of 14 or 16 days, respectively, the increase of the mean values is markedly higher for the left and right ventricle in the experimental group than in the corresponding control group. A similar increase in the mean value of the numbers of cells takes place in the interventricular myocardium of the embryonic heart after a period of 18 to 20 days of incubation in the experimental group, whereas this increase is relatively small in the corresponding control group.

Table 1: Morphometry of the numbers of cells in the compact myocardium of the embryonic chicken hearts (n = 33) in three distinct locations (see fig. 1) correlated to the period of incubation and to the diminished (15 %) or normal (21 %) oxygen supply in the incubation period from day 6 to day 12.

Period of incubation (days):	Incubation: Control (i.e. normoxic) or ex-perimental (i.e. hypoxic from day 6 to day 12)	Location: Left ventricle	Location: Interventricular septum	Location: Right ventricle
Day 10/12 (n = 16)	Control (n = 7)	Min. 98 Mean142.6 Max. 168	Min. 125 Mean 144.8 Max. 164	Min. 119 Mean 140.3 Max. 164
	Experimental (n = 9)	Min. 114 Mean 137.5 Max. 156	Min. 128 Mean 152.2 Max. 174	Min. 113 Mean 142.1 Max. 163
Day 14/16 (n = 9)	Control (n = 4)	Min. 142 Mean 158.5 Max. 176	Min. 138 Mean 152.7 Max. 173	Min. 108 Mean 135.5 Max. 167
	Experimental (n = 5)	Min. 158 Mean 188.4 Max. 199	Min. 144 Mean 166.0 Max. 186	Min. 153 Mean 165.4 Max. 178
Day 18/20 (n = 8)	Control (n = 4)	Min. 138 Mean 161.5 Max. 185	Min. 162 Mean 168.2 Max. 185	Min. 108 Mean 115.7 Max. 129
	Experimental (n = 4)	Min. 140 Mean 164.0 Max. 186	Min. 164 Mean 198.5 Max. 224	Min. 103 Mean 117.2 Max. 128

Abbreviations: Min. = Minimum, Max. = Maximum, Mean: Mean value (arithmetic)

Table 2: Comparison of the mean values of the cell numbers in the myocardium in the control group (normoxic incubation, n = 15) and experimental group (hypoxic incubation from day 6 to day 12, n = 18) correlated to the three distinct locations of the hearts (see fig. 1) and to the period of incubation of the chicken embryos.

	Location: Left ventricle		Location: Interventricular septum		Location: Right ventricle	
Period of incubation	Control	Experimental	Control	Experimental	Control	Experimental
Day 10/12	142.6	137.5	144.8	152.2	140.3	142.1
Day 14/16	158.5	188.4	152.7	166.0	135.5	165.4
Day 18/20	161.5	164.0	168.2	198.5	115.7	117.2

In the control group, the mean value of the numbers of cells of compact myocardium in the outer wall of the left ventricle increases slightly and constantly over the incubation period from day 10 to day 20, whereas in the experimental group the mean value of the cell numbers at first decreases slightly and afterwards increases markedly in the incubation period of day 14/16. The mean values of the numbers of myocardial cells in the interventricular septum of the control group go up slightly from day 10 to day 20 of incubation, but in the control group this increase in the mean value is significantly higher in the incubation period from day 14/16 to day 18/20 (p = 0.015). For the myocardium of the outer wall of the right ventricle the morphometric analysis reveals a slight but constant decrease of the mean value of the cell numbers over the incubation period in the control group, whereas in the experimental group the mean value of the numbers of myocardial cells increases to day 14/16 of incubation, followed by a marked decrease of the cell numbers to the end of the incubation period reaching the level of the control group.

The increase of the mean number of cells in the myocardium of the outer wall of the left and right ventricle occurs in the embryonic hearts in the experimental group after a period of 14 to 16 days of incubation, whereas this significant increase takes place in the interventricular septum after 18 to 20 days of incubation, just before hatching.

The chick embryos incubated under reduced environmental oxygen supply displayed an increase of the mean value cell number regarding the three different locations within the heart (left ventricle, interventricular septum, and right ventricle) compared to the respective locations within the control group. Although these results were only statistically significant for the interventricular septum, they indicate that the reduced environmental oxygen supply indeed affects the structural development and differentiation of the embryonic myocardium. This adaptive mechanism results presumably from a "reactive" hypertrophy of the myocardial cells, as well as from an increased mitotic myocardiac activity, i.e. hyperplasia, during a "sensitive phase" of the chick embryo between the sixth and twelfth day of incubation, respectively.

This adaptive potential of the embryonic myocardial cells – like other parameters described in the literature – is detected only temporarily, in a distinct developmental phase (day 6 to day 12). The hearts of older chicks shortly before hatching of both, the experimental group, i.e. hypoxia, and the control group, i.e. normoxia, displayed nearly identical mean values of cell numbers within the selected locations of the myocardium in the outer ventricular walls. Thus, the detected altered morphometrical and morphological parameters within the experimental group were almost balanced towards the end of the prenatal development (Graphs 1 to 3).

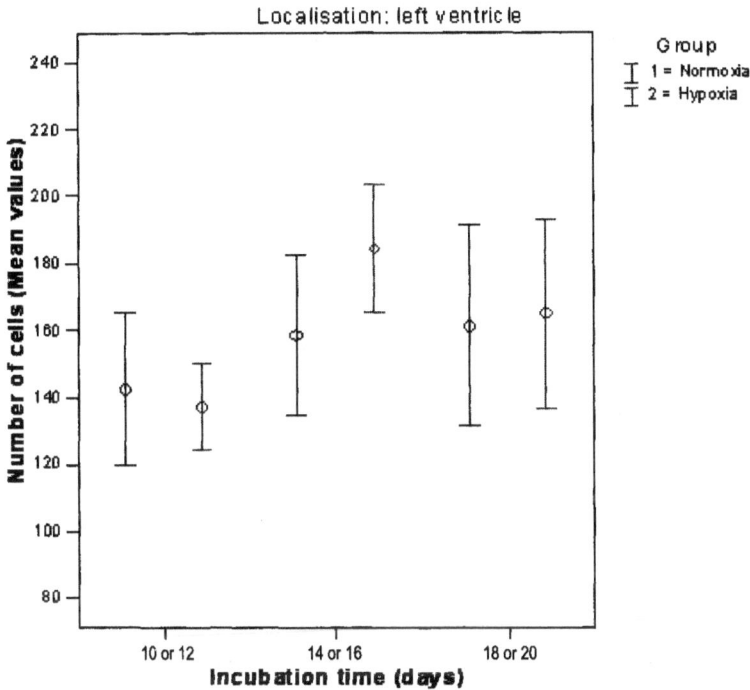

Graph 1: Numbers of cells in the myocardium of the outer wall of the left ventricle in the control group (1 = Normoxia) and in the experimental group (2 = Hypoxia):

The mean value of the cell numbers in the myocardium of outer wall of the left ventricle (see fig. 1) is lower in the first of the experimental group (day 10 or 12 of incubation) than in the corresponding control group, but shows a significant increase in the second group (day 14 or 16 of incubation) to reach a higher level than in the control group of the same incubation period. Towards the end of incubation (day 18 or 20 of incubation), the mean value of the cell numbers is almost balanced between the experimental group and the control group.

Discussion

The physiological prenatal embryonic and foetal growth as well as the maturation of the heart is for the most part a result of hyperplasia, i.e. the mitotic activity of cardiomyocytes (Sedmera *et al.*, 1997). This mitotic activity of the prenatal myocardium is limited. As described by Armstrong *et al.* (2000) the defined and required number of cell divisions is limited for the myocardial tissue. The basic question of this study was whether the reduced environmental oxygen content, i.e. 15 % instead of 21 % oxygen from day 6 to day 12 of incubation, is able to induce an early, i.e. premature, increase of cell divisons resulting in an increase of numbers of cells in the myocardium. The embryonic heart adapts to different functional requirements by modifying its proliferational processes (Sedmera *et al.* 1999), as demonstrated by the results of our study. Prenatally, differences in cell size are less significant in the myocardium (Bical *et al.*, 1990),

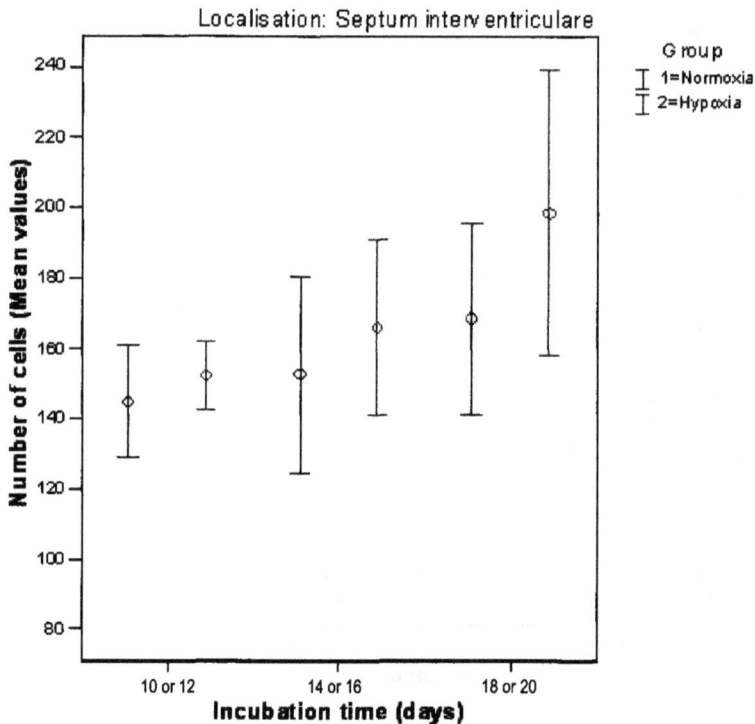

Graph 2: Numbers of cells in the myocardium of the interventricular septum in the control group (1 = Normoxia) and in the experimental group (2 = Hypoxia):

In the interventricular septum, the experimental group shows a significant increase in the mean value of cell numbers in all embryonic stages. Compared to the results of the left and of the right ventricle, a distinct difference of the numbers of cells in the myocardium is detectable between the two groups that increases even significantly to the end of the incubation period near hatching.

indicating that before hatching the adaptation of the myocardium to environmental hypoxic conditions is mostly established by hyperplasia. On the other hand, there is evidence that postnatally the cellular hypertrophy plays a key-role in the adaptive mechanism.

Studies on cultured embryonic cardiomyocytes showed that the myocardial tissue is capable to regulate its own mitotic activity according to the impact of external mechanical forces (Miller *et al.*, 2000). Additionally, Saiki (1997) described the prenatal increase of the proliferation of cardiomyocytes as the consequence of experimentally increased intraventricular pressure. Due to the differences in the ventricular pressure load, i.e. the resistance of the circulation in the body exceeds that in the circulation in the developing lungs (and air sacs), and cardiac efficiency, a region-specific examination of the myocardial sections of the embryonic heart was therefore carried out in the present study. We detected evidence for an increased mean value of cell numbers in the experimental group, whereas Villamor *et al.*

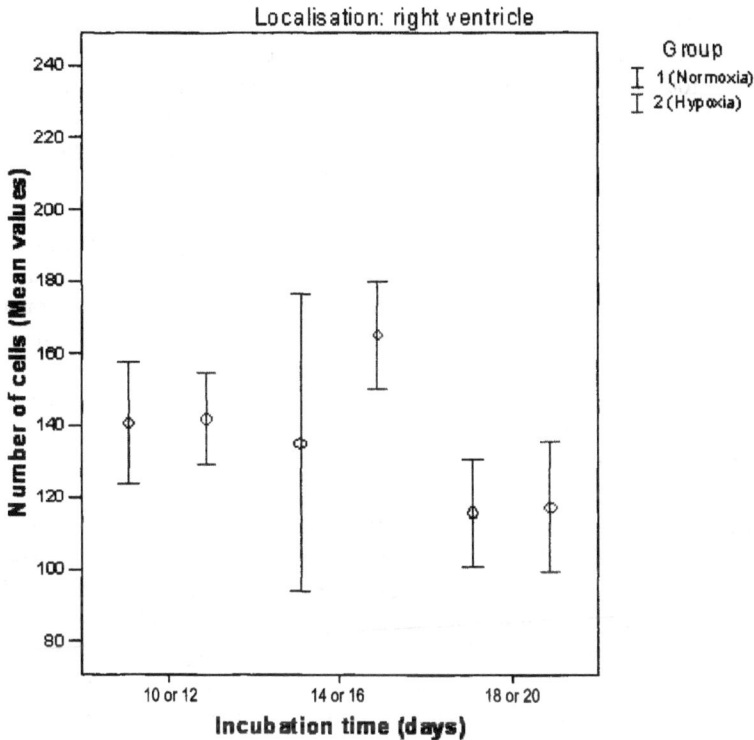

Graph 3: Numbers of cells in the myocardium of the outer wall of the right ventricle in the control group (1 = Normoxia) and in the experimental group (2 = Hypoxia):

Compared to the results of the left ventricle (see Graph 1), an increase in the mean value of the cell number was also detectable in the outer wall of the right ventricle (see fig. 1) at day 14 or day 16 of incubation within the experimental group compared to the control group. The numbers of myocardial cells showed a significant broad range in the control group at day 14 or day 16, respectively, compared to the experimental group. Before hatching, the mean values and ranges of the numbers of cells in the myocardium of the outer wall of the right ventricle diminish and are almost the same in both, the control and the experimental group.

(2004) did not detect any differences in the number of cells in the myocardium of embryos between day 6 and day 19 of hypoxic incubation, but describe a biventricular cardiac enlargement as a consequence of the hypoxic incubation.

In the left ventricle, our results show a slight decrease in the mean value of numbers of myocardiac cells after 10 or 12 days of incubation in the experimental group compared to the control group. This decrease might be due to the direct effect of the incubation in the reduced environmental oxygen supply in the period from day 6 to day 12, thus confirming the finding of Asson-Batres *et al.* (1989). The authors describe a proportional correlation between the oxygen supply and the organic tissue growth. This result might also be – as described in the literature (Dzialowski *et al.*, 2002; Dusseau *et al.*, 1989) – a consequence of the higher sensitiviy of the cardiac tissue to the oxygen supply within the specific phase of the embryonic development of the heart between day 6 and day 12 of incubation. We suspected the myocardiac tissue to be even more susceptible to a reduced oxygen supply during this time frame. The distinct increase in the mean value of cell numbers for the experimental group after 14 or 16 days of incubation is considered to be a "reactive hyperplasia" of the myocardium. This adaptive reaction of the myocardium might be the reaction to the reduced contractive efficiency of the cardiac muscle during the previous "sensitive phase", i.e., the days 6 to 12, therefore the temporal cardiac "insufficiency" is compensated by a higher mitotic activity of the cardiomyocytes, i.e. by hyperplasia of the myocardium.

The left ventricle of the embryonic heart is the most stressed part of the heart regarding the relevance for body circulation and its effectiveness. Before hatching and the onset of breathing, the venous blood returning to the heart passes through the oval foramen in the interartrial septum to the left atrium, i.e. bypassing the right ventricle and the lungs as well as the developing air sacks, and passes then via the left atrioventricular ostium into the left ventricle. The myocardium of the outer wall of the left ventricle and of the interventricular septum generates the main force for promoting the blood circulation in the body, whereas the right ventricle before hatching caries blood into the non-inflated lungs and air sacks only.

Due to the adaptional ability of the heart, the differences of the mean values of the cell numbers in the outer wall of the left and right ventricle detected within the experimental groups compared to the control groups is balanced towards the end of incubation. We therefore conclude that the structural differences generated under the influence of the reduced environmental oxygen content within the left and right ventricle are remarkable.

Other investigators described a similar compensation between the physiologic parameters studied in control and in experimental groups by the time of hatching (Decker, 2002; Dzialowski *et al.*, 2002).

For the interventricular septum of the heart, the mean value of cell numbers in the experimental group is significantly ($p = 0.015$) higher regarding the overall developmental stages (see Graph 2). Towards the end of incubation, after 18 or 20 days of incubation, we detected the most obvious difference between the two groups for the interventricular septum.

We consider the differing region-specific myocardiac differentiation to be the result of the temporally deviating adaptive processes. As described in the literature (Jeter *et al.*, 1971; Tokayusu, 1990; Franco *et al.*, 1997) the interventricular septum is the least formed and remodelled in the course of embryonic development. Compared to the interventricular septum, the outer walls of the ventricles display a trabecular structure in the early stage of embryogenesis. The cardiomyocytes of the trabecular part of the ventricular myocardium (see fig. 2 a) are more differentiated than those of the compact myocardium (Tokayusu, 1990; Franco *et al.*, 1997). Sedmera *et al.* (1997) concluded from this higher degree of differentiation that this mechanism plays a role in the early contractive function of the heart muscle. The advanced differentiated trabecular parts of the ventricular walls might therefore react earlier to a diminished incubational oxygen supply than the myocardium in the interventricular septum. This part of the embryonic myocardium has a compact structure in all stages of prenatal development. For this reason we suggest a delayed foetal development of the septal myocardial cells compared to those in outer ventricular walls. Because of its differing structure, it seems that the cardiomyocytes of the interventricular septum are at the beginning (day 10/12) less sensitive to a reduction in oxygen supply and therefore display minor compensatory mechanisms. But at the end of the prenatal development, the interventricular myocardium proliferates in the experimental group significantly more than the myo-cardium in the outer walls of the ventricles, that show almost the same mean values of cell numbers as in the control group. Our results indicate a regionally different effect on the mean values of cell numbers in the myocardium to the reduced oxygen supply in incubation. This effect of reduced environmental oxygen content is not only dependent on the intensity and the time course of the hypoxic incubation but also on the respective cardiac tolerance (Ostadal *et al.*, 1999). During the embryonic development the authors noticed a significant alteration of this cardiac tolerance. The results of the present study indicate that there might be even differences in the tolerance of the myocardium for oxygen deficiency within the different regions of the heart.

Similar to the left ventricle, we detected a distinct difference in the mean value of cell numbers between the two groups, i.e. experimental and control, after 14 or 16 days of incubation in the outer wall of the right ventricle (see Graph 3). Although there was no decrease in the number of cells after 10 or 12 days of incubation – likewise comparable to the left ventricle – we suggest a "reactive myocardiac hyperplasia" in the right ventricle after 14 or 16 days of incubation. The mean values of the numbers of myocardial cells in the outer wall of the right ventricle in both, the experimental and in the control group, decrease in the last period of incubation. This result may be explained by the end of the hyperplasia and onset of the hypertrophy in the myocardial cells prenatally compared to the cells in the outer wall of the left ventricle and especially compared to the cells in the interventricular septum. In both of these locations, the hypertrophy may start after hatching and may then replace the hyper-plasia in this part of the myocardium.

Besides the local effects of the reduced oxygen supply of the embryo in its egg directly related to the development of the myocardium, systemic effects of the reduced oxygen supply on the circulating blood, i.e. an increased blood volume or an increase in the number of erythrocytes, may explain the various structural changes in the embryonic myocardium of the chicken. Additionally, a varying development of the myocardial blood supply in the three different locations of the ventricles may - to a different degree - counterbalance the influence of a diminished oxygen supply, thus explaining the varying responses of the myo-cardium to the reduced oxygen supply in incubation. Further investigations, particularly electron microscopy, will be necessary to gain more information on the cardiomyocyte struc-ture and its developmental changes as well as on its blood supply within the different re-gions of the heart.

Acknowledgement

The authors would like to thank K. Briest-Forch and I. Küster-Krehahn for excellent techni-cal support. Further we would like to thank G. Arndt and K. Hornemann (Institut für Biometrie und Informationsverarbeitung, FU Berlin) for supporting the statistical evaluation.

This study is financially supported by the Senate of Berlin (NaFöG-stipend).

References

ARMSTRONG, M. T., D. Y. LEE, and P. B. ARMSTRONG (2000): Regulation of Prolif-eration of the Fetal Myocardium. *Dev. Dyn.* **219**, 226-236

ASSON-BATRES, M. A., M. K. STOCK, J. F. HARE, and J. METCALFE (1989): O$_2$

effect on composition of chick embryonic heart and brain. *Resp. Physiol.* **77**, 101-110

BICAL, O., P. GALLIX, M. TOUSSAINT, P. LANDAIS, D. GAILLARD, J. KARAM, and J. Y. NEVEUX (1990): Intrauterin versus postnatal repair of created pulmonary artery stenosis in the lamb. Morphologic comparison. *J. Thorac. Cardiovasc. Surg.*, **99**, 685-690

DECKER, S. (2002): Der langfristige Einfluss eines verminderten O_2-Angebotes während der Inkubation auf Katecholamine, Stoffwechselmetabolite sowie Hämatokrit und Hämoglobin im Blut von Hühner- und Entenembryonen. [Long term influence of reduced oxygen supply on catecholamines, metabolic parameters as well as hematocrit and hemoglobin in chicken (Gallus gallus f. domestica) and duck (Cairina moschata) embryos.] Berlin, Freie Universität Berlin, Faculty of Veterinary Medicine, Dissertation Thesis

DUSSEAU, J. W., and P. M. HUTCHINS (1989): Microvascular responses to chronic hypoxia by the chick chorioallantoic membrane: a morphometric analysis. *Microvasc. Res.*, **37**, 138-147

DZIALOWSKI, E., D. von PLETTENBERG, N. A. ELMONOUFY, and W. W. BURGGREN (2002): Chronic hypoxia alters the physiological and morphological trajectories of developing chicken embryos. *Comp. Biochem. Physiol. A.*, **131**, 713-724

FRANCO, D., J. YA, G. T. M. WAGANAAR, W. H. LAMERS, and A. F. M. MOORMAN(1997):The trabecular component of the embryonic ventricle. In: Ostadal, B., M. Nagano, N. Takeda, N. S. Dhalla (eds.): *The developing heart.* Philadelphia, Lippincott – Raven, Pp. 51-60

HÜHNKE, A. (2003):Der Einfluss von Sauerstoffmangel in der Inkubationsluft auf den Blut-Gas-Status und morphologische Parameter von Hühnerembryonen *(Gallus gallus f. domestica).* [The influence of a reduced oxygen content in the incubation air on the blood gas status in the chorioallantoic vein and morphologic parameters in the chicken embryo *(Gallus gallus f. domestica).*] Berlin, Freie Universität Berlin, Faculty of Veterinary Medicine, Dissertation Thesis

JETER, J. R., and I. L. CAMERON (1971): Cell proliferation patterns during cytodifferentiation in embryonic chick tissues: liver, heart and erythrocytes. *J. Embryol. Exp. Morph.*, **23**, 405-422

MILLER, C. E., K. J. DONLON, L. TOIA, C. L. WONG, and P. R. CHESS (2000): Cyclic strain induces proliferation of cultured embryonic heart cells. *In vitro Cell Dev. Biol. Anim.*, **36**, 633-639

OSTADAL, B, I. OSTADALOVA, and N. S. DHALLA (1999): Development of cardiac sensitivity to oxygen deficiency: comparative and ontogenetic aspects. *Physiol Rev.*, **79** (3), 635-659

ROMEIS, B. (1989): Mikroskopische Technik, 17. Aufl. München, Urban u. Schwarzenberg

SAIKI, Y., A. KÖNIG, J. WADDELL, and I. M. REBEYKA (1997): Hemodynamic alteration by fetal surgery accelerates myocyte proliferation in fetal guinea pig hearts. *Surgery,* **122**, 412-419

SEDMERA, D., T. PEXIEDER, V. RYCHTEROVA, N. HU, and E. B. CLARK (1997): Developmental Changes in the myocardial architecture of the chick. *Anat. Rec.*, **248**, 421-432

SEDMERA, D., T. PEXIEDER, V. RYCHTEROVA, N. HU and E. B. CLARK (1999): Remodeling of Chick Embryonic Ventricular Myoarchitecture Under Experimentally Changed Loading Conditions. *Anat. Rec.*, **254**, 238-252

TOKUYASU, K. T. (1990): Co-development of embryonic myocardium and myocardial circulation. In: E. B. Clark, A. Takao, (eds.) "Developmental Cardiology: Morphogenesis and function". New York, Futura, Pp. 205-218

VILLAMOR, E., C. G. A. KESSELS, K. RUITJENBEEK, R. J. VAN SUYLEN, J. BELIK, J. G. R. DE MEY, and C. E. BLANCO (2004): Chronic in ovo hypoxia decreases pulmonary arterial contractile reactivity and induces biventricular cardiac enlargement in the chicken embryo. *Am J. Physiol. Regul. Integr. Comp. Physiol.*, **287** (3), 642-651

Video analysis of body movements and their relation to the heart rate fluctuations in chicken hatchlings

YONETA, H., AKIYAMA, R., NAKATA, W., MORIYA, K.[1] AND TAZAWA, H*.

Department of Electrical and Electronic Engineering, Muroran Institute of Technology, Muroran 050-8585; [1]Department of Electrical and Electronic Engineering, Hakodate National College of Technology, Hakodate 042-8501, Japan.

In order to elucidate relationship between body movements of chicks and particular fluctuations of instantaneous heart rate (IHR) which have been categorized into three types (Types I, II and III), we developed an image processing system to quantify the amount of movements of the wings and the whole body. A CCD camera captured movements of the wing or the body into a computer. For analysis of wing movements, wings were coloured in red to distinguish from other parts of the body. For the whole body analysis, the green component of the colour image was extracted from the video image data and processed for the amount of movements. Using the image processing system developed, we made simultaneous determination of IHR from electrocardiogram and amount of movements of the wing or the whole body. We found that Type I and Type II HR oscillations were related to the periodic movements of the wing and Type III HR irregularities occurred simultaneously with the spontaneous whole body movements; a stagger or twitch. As a result, it was confirmed that the chick moved the body at the same frequencies as Type I and II HR oscillations and simultaneously with Type III HR irregularities. These HR fluctuations and body movements may be attributed to the same origins.

Keywords: heart rate fluctuations, chicken hatchling, CCD camera, body movement, wing movement; difference; binarization

Introduction

In newly hatched chicks, instantaneous heart rate (IHR) fluctuates cyclically and/or irregularly and IHR fluctuations have been categorized into three types; that is, Types I, II and III (Moriya et al., 1999). Type I and Type II heart rate (HR) fluctuations are heart rate variability (HRV) which oscillates at a mean frequency of about 0.7 Hz (i.e., high frequency oscillation) and 0.07 Hz (low frequency oscillation), respectively. Type III HR fluctuation is heart rate irregularities (HRI) which are dominated by non-cyclic spontaneous HR accelerations. By means of simultaneous measurements of IHR and respiratory signal, Type I HRV was shown to be respiratory sinus arrhythmia (RSA) (Moriya et al., 2003). Type II HRV was often produced by exposing the hatchlings to a low ambient temperature (Tazawa et al., 2001, 2002). Type III HRI was

*Corresponding author: TAZAWA, Hiroshi
E-mail: tazawa@mmm.muroran-it.ac.jp

often recorded when the head of hatchlings moved spontaneously (Moriya *et al.*, 2003). Accordingly, it is likely that these three types of HR fluctuations occur together with body movements of hatchlings. It remains to be investigated how hatchlings move and what kind of body movements are related to individual HR fluctuations. We assume that the wings move periodically, which is related to Type I and Type II HR oscillations, and the whole body moves spontaneously, which is related to spontaneous Type III HR irregularities. In the present study, first we develop an image processing system to capture movements of the wings or the whole body into a computer using a CCD (Charge Coupled Device) camera and quantify the movements by difference method and binarization of pixels comprising particularly processed images. Using the image processing system with an electrocardiogram, we determine simultaneously IHR and the amount of movements of the wing or the body and elucidate relationship between them.

Materials and methods

CHICKS AND INSTANTANEOUS HEART RATE

Fertile eggs of broiler chickens were incubated at 38°C and moved to a hatching tray on the bottom of the incubator on day 20 of incubation. When chicks hatched, they were transferred to a brooding cage which was put in a laboratory room. Ambient temperature (Ta) was about 24-27°C. Food and water were given *ad libitum*. Chicks were brooded there until the experiment was performed on Days 4-8 after hatching. On the day of experiment, a chick was fitted with three flexible electrodes of electrocardiogram (ECG) on the lateral thoracic wall under both the wings and on the ventral abdomen as mentioned elsewhere (Moriya *et al.*, 1999, 2000). IHR was calculated from time interval of adjacent R waves which crossed a threshold set on the computer (Moriya *et al.*, 1999, 2000). Because time intervals between individual values of IHR were not equal, IHR at an equal interval (i.e., 1/15 sec) was calculated by a linear interpolation for power spectrum analysis.

EXPERIMENTAL CHAMBER

The chick with ECG electrodes was placed in an experimental chamber (20 x 30 x 20 cm). The outside of the chamber consisted of a metal-mesh box grounded to attenuate external interference noises, and a cardboard box painted in black was placed inside. The recess of the box was formed by a triangular pillar to attenuate diffused reflection of light to the camera. The chick was illuminated indirectly to reduce sharp contrast between light and dark. The front of the chamber was made a peephole for a CCD camera (Type MTV-5366ND, Akizuki Denshi Tsusho Co. Ltd, Tokyo, capturing rate; 30 frames/sec, effective pixels; 768 x 494, lens diameter; 14.0 mm, colour; 24-bit) which captured an image of the chick into a computer at a rate of 30 frames/sec. For experiment in relation to Type II HRV, the chamber having the chick in it was placed in the laboratory at 24-27°C, to increase the chance of its occurrence. For experiment related to Type I and Type III HR fluctuations, the chamber was put in a large incubator warmed at 35°C to make the chick relax.

QUANTIFICATION OF WING MOVEMENT

Basically, quantification of wing movement was based on taking difference of pixels between adjacent frames of the video-image of the wing to produce a difference frame, and the amount of wing movement was determined by the total number of pixels in the difference frame after binarization. Both wings were coloured in red so that they could be distinguished clearly from other parts of the body for their abstraction. The video-image of coloured wings was captured by the CCD camera in a MPEG (Motion Picture Experts Group) format according to specifications for the capture board of the CCD camera. Then, MPEG image was transformed into the AVI (Audio Video Interleaving) image in order to take the difference between the consecutive frames with gradations. The transform from MPEG to AVI image was made every two frames in order to reduce the number of steps of the difference procedure; that is, 15 frames/sec.

The individual frames of the AVI image were differentiated into three colour elements; red, green and blue in 8-bit gradations (0-255), respectively, and divided into three pictures that were each composed of one of the three elements. Because these three pictures were depicted by gradations of pixels of red, green or blue, the gradation range of pixels which composed preferentially the wing was assigned to individual colour elements. These gradation ranges for red, green and blue pixels made an indispensable condition to abstract the wing from the AVI image. The individual pixels comprising the AVI image were examined for whether they satisfied the condition; that is, gradations of a given colour pixel were within the three gradation ranges simultaneously. Then, the pixels which satisfied the condition were allotted *1* and white colour in an abstract frame, otherwise *0* and black colour (i.e., binarization) in order to abstract the wings from other parts of the body. Thus, the image of wings was depicted in the abstract frame. Finally, the difference of abstracted wings between adjacent two frames was taken and binarization of pixels was made to quantify the amount of wing movement. The pixels which changed between the two adjacent frames were allotted *1* and white colour in a new difference frame, and those without change, *0*. Total number of pixels allotted *1* in the difference frame determined the amount of wing movement. Accordingly, the amount of wing movement was expressed by the number of pixels.

QUANTIFICATION OF WHOLE BODY MOVEMENT

Basically, quantification of whole body movement was similar to that of wing movement based on the difference method and binarization of pixels comprising the difference image. Capture of whole body movement by the CCD camera into the computer was the same as that of wing movement except colour painting. Colour painting of the body was not needed for abstract of whole body movement, because the body itself could be distinguished from the environment. In this context, fundamentally the abstract of whole body movement could be made from images captured by a monochrome camera. Because a simple difference between adjacent two monochrome images was disturbed by flickers due to thermal noises, adverse effects by flickers were eliminated by setting a threshold with subsequent binarization of pixels. The pixels exceeding over the threshold were allotted *1* and white colour in a new difference frame, otherwise *0* and black colour. This procedure depicted moving part of the body in white and motionless part in black. The total number of white pixels in the difference frame was counted as the amount of body movement.

The employment of a colour CCD camera made it possible to determine simultaneous quantification of movements of the wing and the whole body. However, the additional image processing was needed; that is, to differentiate the colour AVI image of the whole body into three colour components; i.e., red, green and blue. The blue image was contaminated with noises, and either the red or green image which was less contaminated could be used for quantification of whole body movement. Because partition of the green element for transform from colour to monochrome was larger than the red element, the green image was used in the present system. The difference between adjacent two green frames was taken to produce a difference frame. In this difference frame, not only body movement, but also noises caused by diffused reflection of light were included. Because the green pixels comprising body movement had high gradations and the gradations of diffused reflection noises were low, the threshold was set at 40. The green pixels over 40 were allotted *1* and white colour in a binary frame, otherwise *0* and black colour. The total number of *1* in the binary frame was counted as the amount of the whole body movement this instant.

POWER SPECTRUM ANALYSIS

In order to confirm whether IHR and movement of the wing or the whole body were periodic and to determine a frequency of periodicity, power spectrum analysis (PSA) was made for IHR and movement of the wing or the body. IHR data interpolated at 1/15 sec and somatic movements represented by the number of pixels every 1/15 sec were divided every 1024 points, first from the beginning of the given recordings and then from the first 512 points of the same recordings in order to increase the number of sections analyzed. The power spectrum density (PSD) of frequencies in the recording comprising each series of 1024 points was then calculated by fast Fourier transform (FFT) using Hamming windows (Harris, 1978). The spectra analyzed were averaged in order to improve resolutions of the spectrum peak.

Results

THE AMOUNT OF WING MOVEMENT

Figure 1 illustrates an abstraction of the wing movement and quantification of its amount. The top frame shows a single AVI image of a chick whose wings were coloured in red. Three pictures in the second line are composed of one of three colour elements; that is, red, green and blue from *left*, respectively. For these pictures, the gradation range of pixels which depicted preferentially the wing was assigned as in the third line. In this example, the gradation range was assigned to be 90-160, 40-80 and 40-100 for red, green and blue elements, respectively. Then, individual pixels in the AVI image (top frame) were examined whether they satisfied these three gradation ranges. The pixels which satisfied them were allotted white colour to depict the wings to be abstracted. The abstracted image of the wing is shown on the *left* of the fourth line. On the *right* is shown the image of the wing which was abstracted from the next AVI image 1/15 sec later. The difference of abstract image of the wing between these two adjacent frames was taken, and the pixels which changed between these two frames were allotted *1* and white colour in a new difference image (bottom line). The total number of white pixels in the difference image was designated as the amount of wing movement; that is, 1,286 pixels for this frame. In like manner, the wing movement was quantified 15 times during 1-sec period.

Relationship between Type I heart rate oscillation and wing movement: Figure 2 presents a simultaneous measurement of IHR and wing movement and their PSD. In the 5-min

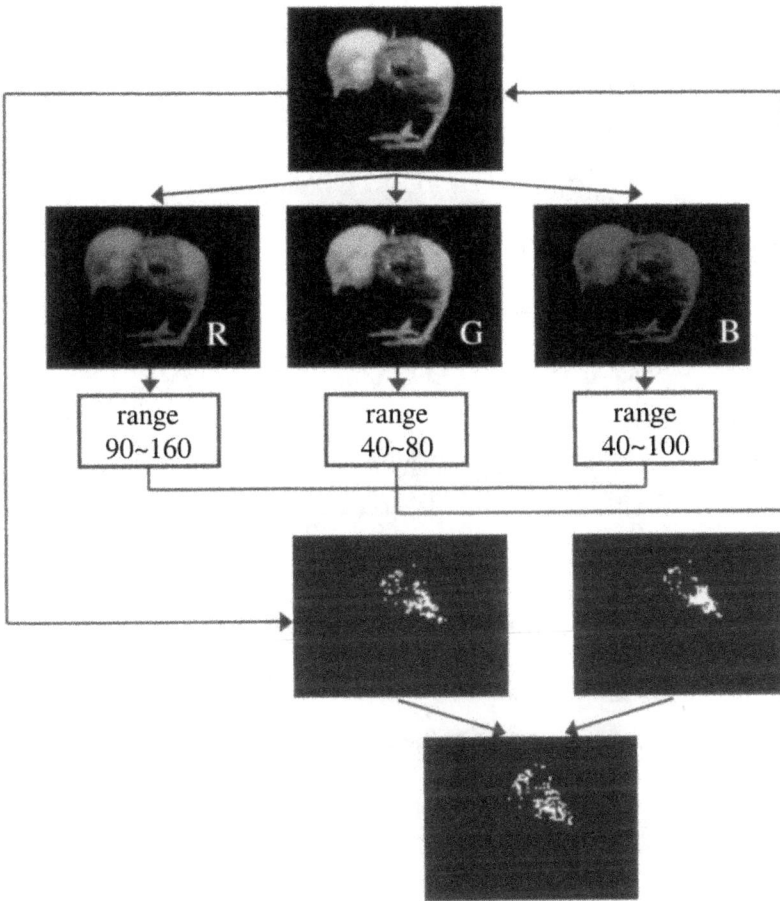

Figure 1. The abstraction of a wing movement and quantification of its amount in a 7-day-old chick which was set electrocardiogram electrodes under the wings. The binary difference image of the wing in the bottom panel shows the wing movement during the period of 1/15 sec.

recording (top panel), IHR baseline was wide with a range of about 30 bpm (top tracing) and no particular periodicity was seen in the wing movement (bottom tracing). In the time expanded 30-sec recording (middle panel), it is shown that IHR oscillated (top tracing) and the wing movement most likely oscillated (bottom tracing). PSA shows that both IHR and wing movement had a peak of PSD at the frequency of 0.62 Hz (bottom panel). The oscillation of IHR was Type I HRV due to RSA and coincided with the wing movement. As a result, breathing of the chicks moved the wing and thus their respiratory movements could be detected from the wings. Five chicks were subjected to the simultaneous measurement of Type I HR oscillation (RSA) and wing movement. The mean frequency of RSA was 0.73± 0.12 (SD) Hz.

    Relationship between Type II heart rate oscillation and wing movement: Figure 3 shows a simultaneous measurement of IHR and wing movement and their PSD. IHR in the top tracing oscillated with a low frequency; i.e., Type II HR oscillation (top panel). PSA indicates that the oscillating frequency was 0.089 Hz. Because the wing movement in the middle

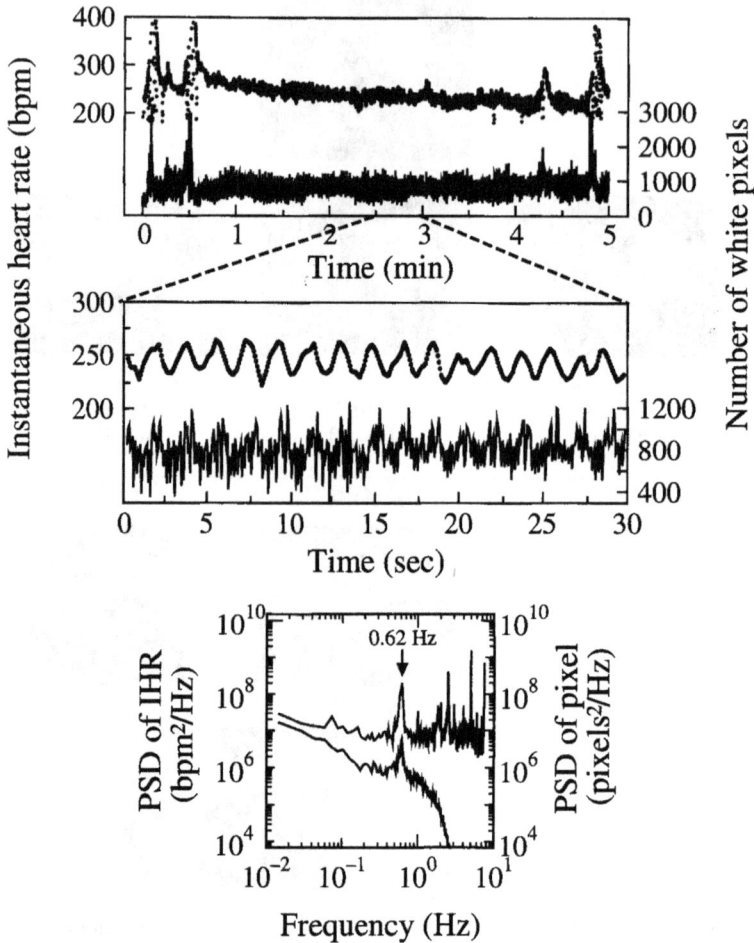

Figure 2. The simultaneous measurement of instantaneous heart rate (IHR) and wing movement of an 8-day-old chick and the power spectrum density of frequency comprising 5-min recording of IHR and wing movement. Top panel; 5-min recording of IHR (top tracing in black) and wing movement depicted by number of pixels (bottom tracing in grey). IHR spontaneously accelerated during the early and late periods of the recording. Middle panel; 30-sec recording that was expanded from the 5-min recording. The ordinate was also expanded. Because the wing movement was presented by 450 points every 1/15 sec, interpolation of IHR was made at the same rate and thus IHR was depicted by 450 points. Bottom panel; power spectrum density (PSD) of IHR (bottom tracing in black) and wing movement (top tracing in grey). The peak of PSD was found at a frequency of 0.62 Hz for both IHR and wing movement.

tracing had high frequency components due to breathing, it was low-pass filtered by a digital filter cutting out above 0.5Hz to exclude breathing movement. The bottom tracing was the wing movement of which high frequency was filtered out. PSA indicates that the wing moved at the same frequency as Type II HR oscillation (0.089Hz; bottom panel). Five chicks aging 4-8 days were subjected to the simultaneous measurement of Type II HR oscillation and wing movement. The mean of frequency was 0.085 ± 0.020 (SD) Hz.

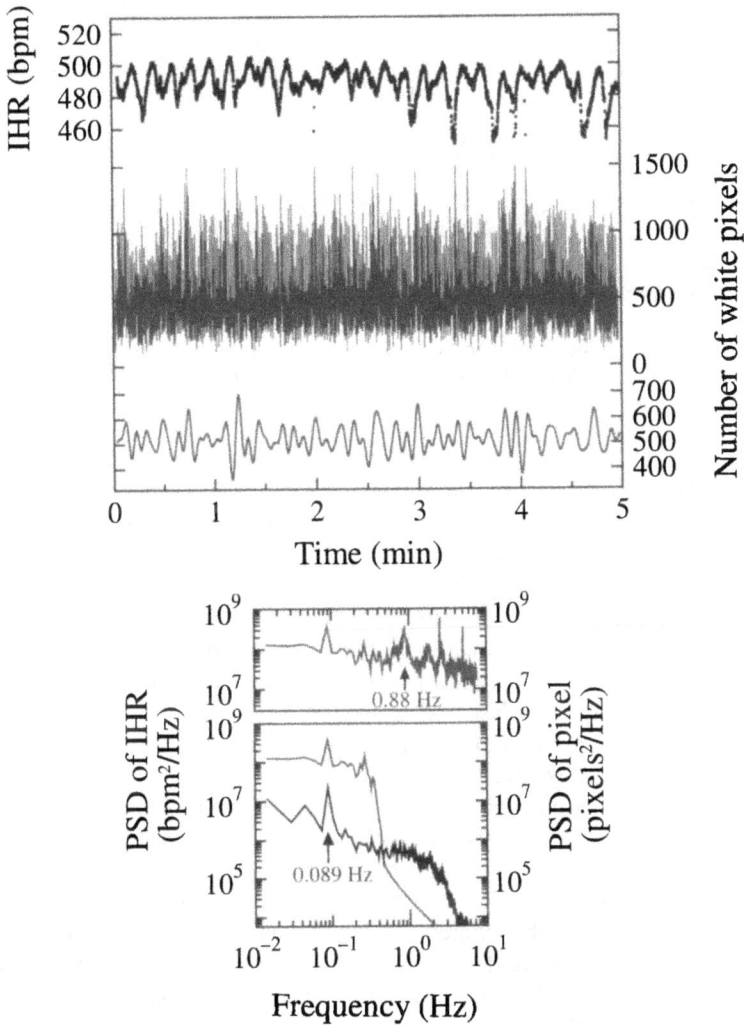

Figure 3. The simultaneous measurement of instantaneous heart rate (IHR) and wing movement of a 7-day-old chick and their power spectrum density (PSD). The measurement was made at 26°C-enviroment. Top panel; 5-min recording of IHR (top tracing in black) and wing movement (middle and bottom tracings in grey). The wing movement in the bottom tracing was low-pass filtered. IHR and wing movement were expanded in magnitude compared with those in Figure 2. Middle panel: power spectrum density (PSD) of wing movement shown in the middle tracing of top panel. PSD had a peak at frequency of 0.88Hz in addition to another peak at 0.089Hz. Bottom panel; PSD of IHR (bottom tracing in black) and low-pass filtered wing movement (top tracing in grey). Both PSD's had a peak at frequency of 0.089Hz.

THE AMOUNT OF WHOLE BODY MOVEMENT

The whole body movement was determined by two ways using either monochrome camera or colour camera. Figure 4 presents determination of the amount of whole body movement

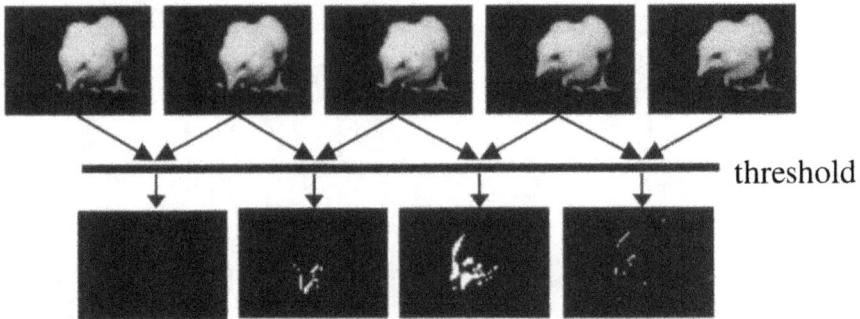

Figure 4. The determination of body movement captured by a monochrome camera during the period of 1/3 sec. The chick was 6-day-old and placed in an environment at 35°C.

in a chick using the monochrome camera (Type IBC-25E, Akizuki Denshi, effective pixels; 512 x 492). The upper five panels show the chick captured at the interval of 1/15 sec during the period of 1/3 sec. The difference between adjacent two frames was taken and the pixels in the difference frame which was disturbed by flickers were compared with the threshold set to minimize the disturbance by flickers. The pixels exceeding over the threshold were allotted white colour in the bottom difference frames. In this example, the chick did not move during the first 1/15 sec-period (left panel) and slightly moved the face around the beak during the next 1/15 sec (second panel) with subsequent large movement around the beak (third panel). During the last 1/15 sec-period, a part of the head slightly moved (right panel). The number of pixels for these three movements corresponded to 950 pixels, 3,100 pixels and 544 pixels, respectively.

Figure 5 shows determination of the amount of whole body movement in a chick employing the colour camera which was also used for the wing movement. The AVI image of the chick (top panel) was differentiated into three images comprising red, green or blue element (second line). Because of low noises in the green image and large partition for transform to monochrome, the difference between adjacent two green images was taken to produce a difference frame (third line). The noises caused by diffused reflection of light in the difference frame (third line) were removed by setting the threshold of 40. The pixels exceeding over the threshold were allotted white colour in the binary frame (bottom line). The total number of white pixels was 1,347 pixels, which represented the amount of body movement in this instance.

Figure 6 shows an example of large body movement corresponding to Type III HR irregularities. The chick sitting on the floor suddenly stood up backward during the period of 200 msec (3/15 sec) (top line). The green image was differentiated individually from the four AVI images (second line). Then, the difference between adjacent two frames was taken (third line). Green pixels exceeding over the threshold of 40 were allotted *1* and white colour in the binary difference frames (bottom line). The number of white pixels was 9,957 pixels, 9,857 pixels and 5,997 pixels for individual binary images from *left*.

Relationship between Type III heart rate irregularities and whole body movement: Figure 7 shows a simultaneous determination of IHR and whole body movement during the 10-min period. IHR baseline was wide due to RSA and increased transiently and spontaneously more than 50 bpm six times during 10-min recording. IHR accelerations that occurred during the 3-4 min period reached 600 bpm and concurrent body movements ex-

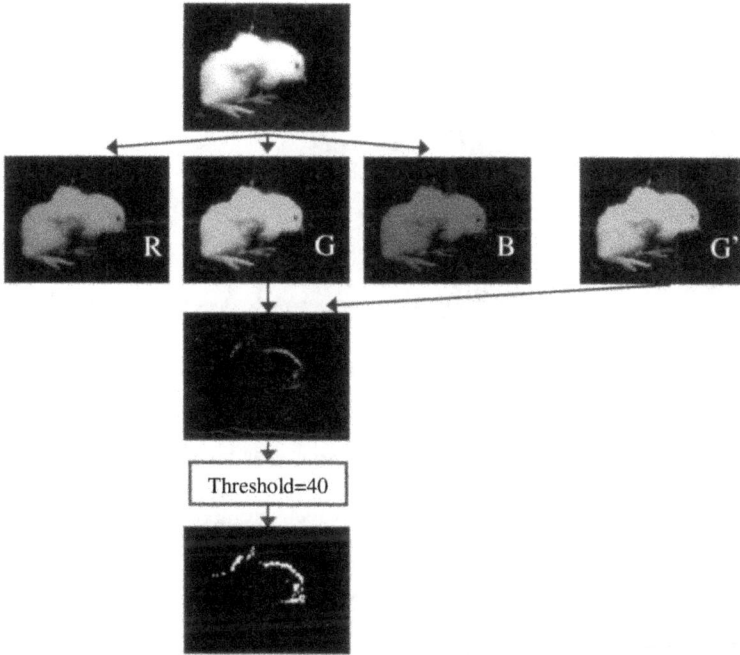

Figure 5. The determination of body movement captured by a colour camera. The chick was 5-day-old and ECG electrodes were attached on the chest wall under the wing.

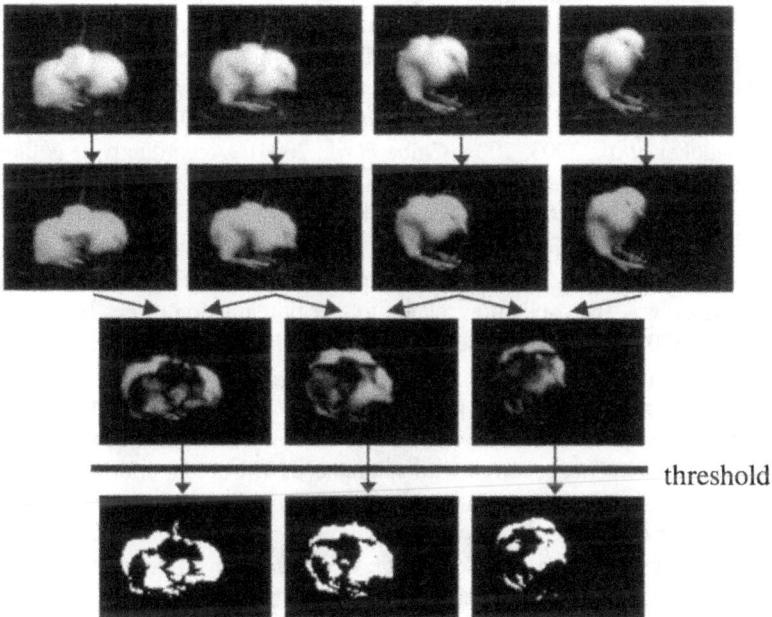

Figure 6. The example of large body movement, which occurred during 200-msec period and corresponded to Type III HR irregularities. The chick was 5-day-old.

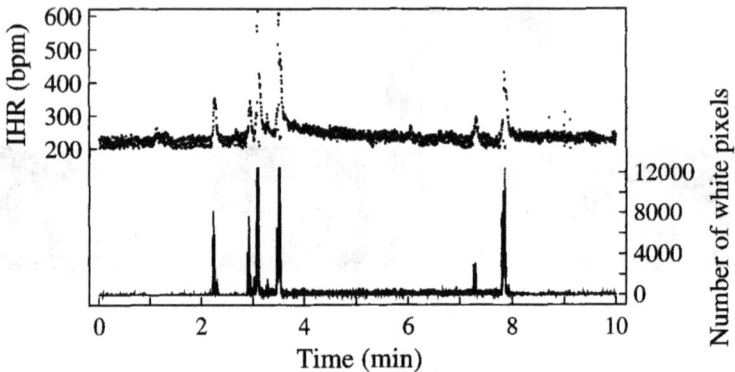

Figure 7. The simultaneous determination of instantaneous heart rate (IHR, top tracing) and whole body movement (bottom tracing). The chick was the same as in Figure 6.

ceeded over 12,000 pixels. Other simultaneous measurements made for nine chicks also showed that chicks staggered or twitched when IHR accelerated largely.

Discussion

Body movements of chicks, which originated in their breathing, could be frequently seen by the naked eye. Visual inspection of AVI images of chicks, which were captured by the CCD camera, certified that the wings moved with breathing in some chicks and twitched several times during a 1-min period in some other chicks. In addition, staggering of chicks was apparent by the eye and confirmed by visual inspection of AVI image. In chickens, both embryos and hatchlings, RSA was frequently recorded and proved by simultaneous measurement with respiratory signal (Moriya *et al.*, 1999, 2000, 2003; Tazawa *et al.*, 1999, 2002b; Khandoker *et al.*, 2003, 2004; Chiba *et al.*, 2004). Accordingly, it is thought that RSA should occur simultaneously with somatic movements caused by breathing. In hatchlings, in addition to RSA which is categorized as Type I HR oscillation, it was reported that low HR oscillation which is categorized as Type II HRV tended to occur in a cool environment (Moriya *et al.*, 1999; Tazawa *et al.*, 2001, 2002a). Thus, it has been assumed that Type II HRV was HR oscillation related to thermoregulation of newly hatched chicks (Tazawa *et al.*, 2002a) and should be accompanied with somatic movements. These assumptions require quantification of somatic movements for comparison with HRV. In addition, Type III HRI has been shown to occur with movement of the head which was detected by a condenser microphone (Moriya *et al.*, 2003). However, it remained to be known what kind of somatic movements occurred with spontaneous HR accelerations.

The present video-image processing system quantified somatic movements of chicks due to the number of pixels. In a cool environment, chicks shivered and wings vibrated minutely. Suppose shivering is minute somatic vibrations with a frequency of several times a second, IHR oscillation corresponding to such somatic vibrations was not yet found in chicks and accordingly quantification of shivering was out of the scope of the present study. The high frequency oscillation of IHR in chicks was 1-2 times a second, caused by breathing. The breathing activity moved the body and accordingly the wings. The movement of the wing was subjected to the present quantification in relation to HRV, because small wings moved frequently more than other parts of the body. Instead, colouring was required to distinguish from the latter. We used innocuous food red.

The movement of the wing was shown by the white pixel in the binary difference image (Figure 1) and the amount of the movement was quantified by the number of pixels. In this instance shown in Figure 1, the number of pixels was 1,286 which corresponded to a large movement of the wing (Figure 2). The number of pixels which was plotted every 1/15 sec simultaneously with interpolated IHR ranged from about 400 to 1,200 pixels. Another example shown in Figure 3 indicates that the wing movement was depicted by pixels with similar number. While the wing movement in Figure 2 was composed of a single oscillation due to breathing, the wing movement in Figure 3 comprised two oscillations with low frequency in addition to high frequency of breathing. The wing moved with the period of 1.1 sec (0.88Hz) due to breathing and additionally with long period of about 11 sec (0.089Hz). This additional slow movement coincided with Type II HR oscillation. Because Type II HR oscillation tended to occur in a cool environment, it is presumed that chicks twitch periodically the body with a period of 10-20 sec in relation to thermoregulation. However, this presumption remains to be studied whether it is related to thermoregulation.

IHR recording in Figure 2 shows that HR baseline transiently accelerated. In accordance with HR accelerations, the movement of the wing increased the number of pixels to more than 3,000 pixels, indicating that spontaneous HR accelerations; i.e., Type III HRI, occurred simultaneously with somatic movements. If a goal of video-image analysis is to quantify the whole body movement, it could be made using the monochrome camera (Figure 4). In order to develop the video analysis to simultaneous determination of movements of both the whole body and the wing, we also used the colour camera to quantify the whole body movement (Figures 5 and 6).

Figures 5 and 6 present an example of small and large body movements, respectively. The small body movements as shown in Figure 5 were not accompanied by pronounced HR accelerations. The large body movements were presented by the large number of pixels (Figure 6) and occurred simultaneously with pronounced HR accelerations (Figure 7). The chick in Figure 6 staggered and HR increased over 400 bpm during a stagger. The heart of the chick beat not less than twice while the chick stood up within 0.2 sec in Figure 6. It is likely that HR accelerations and a stagger were provoked simultaneously by the same origin.

References

CHIBA, Y., YONETA, H., FUKUOKA, S., AKIYAMA, R. and TAZAWA, H. (2004) Ontogeny of respiratory sinus arrhythmia in the domestic fowl. *Avian and Poultry Biology Reviews* **15**: 179-187.

HARRIS, F.J. (1978) On the use of windows for harmonic analysis with the discrete Fourier transform. *Proceedings of the IEEE.* **66**: 51-83.

KHANDOKER, A.H., DZIALOWSKI, E.M., BURGGREN, W.W. and TAZAWA, H. (2003) Cardiac rhythms of late pre-pipped and pipped chick embryos exposed to altered oxygen environments. *Comparative Biochemistry and Physiology* A **136**: 289-299.

KHANDOKER, A.H., FUKAZAWA, K., DZIALOWSKI, E.M., BURGGREN, W.W. and TAZAWA, H. (2004) Maturation of the homeothermic response of heart rate to altered ambient temperature in developing chick hatchlings (*Gallus gallus domesticus*). *American Journal of Physiology. Regul. Integr. Comp. Physiol.* **286**: R129-R137.

MORIYA, K., HÖCHEL, J., PEARSON, J.T. and TAZAWA, H. (1999) Cardiac rhythms in developing chicks. *Comparative Biochemistry and Physiology* **A 124**: 461-468.

MORIYA, K., PEARSON, J.T., BURGGREN, W.W., AR, A. and TAZAWA, H. (2000) Continuous measurements of instantaneous heart rate and its fluctuations before and after hatching in chickens. *Journal of Experimental Biology* **203**: 895-903.

MORIYA, K., CHIBA, Y., YONETA, H., AKIYAMA, R. and TAZAWA, H. (2003) Simultaneous measurements of instantaneous heart rate and breathing activity in newly hatched chicks. *British Poultry Science* **44**: 761-766.

TAZAWA, H., MITSUBAYASHI, H., HIRATA, M., HÖCHEL, J. and PEARSON, J.T. (1999) Cardiac rhythms in chick embryos during hatching. *Comparative Biochemistry and Physiology* **A 124**: 511-521.

TAZAWA, H., MORIYA, K., TAMURA, A., KOMORO, T. and AKIYAMA, R. 2001. Ontogenetic study of thermoregulation in birds. *Journal of Thermal Biology* **26**:281-286.

TAZAWA, H., MORIYA, K. and AKIYAMA, R. (2002a) Low-frequency oscillation of instantaneous heart rate in newly hatched chicks. *Comparative Biochemistry and Physiology* **A 131**: 797-803.

TAZAWA, H., AKIYAMA, R. and MORIYA, K. (2002b) Development of cardiac rhythms in birds. *Comparative Biochemistry and Physiology* **A 132**: 675-689.

Expression of cPer2 and E4bp4 genes in the heart of chicken during embryonic and postembryonic life

I. HERICHOVÁ[1], D. LAMOŠOVÁ[2] AND M. ZEMAN[1,2]

[1]Department of Animal Physiology and Ethology, Comenius University, Mlynska dolina B-2, 842 15 Bratislava, Slovakia; [2]Institute of Animal Biochemistry and Genetics, SASci, 900 28 Ivanka pri Dunaji, Slovakia

Most physiological functions express a rhythmic daily pattern reflecting daily changes in light and temperature and food availability. Information about environmental conditions is mediated to the autonomous circadian oscillator localised in the brain (master oscillator) that drives peripheral oscillators through neural and humoral signals. Peripheral oscillators localised in the heart, liver, pancreas etc. are believed to synchronise local metabolic processes. The aim of our work was to study peripheral clocks in the heart during the prenatal and postnatal life of the chicken. We measured a daily pattern of clock gene *cPer2* and *E4bp4* expression and plasma glucose concentration in 19-day-old embryos and 4-day-old chicks. Expression of clock gene *cPer2* and clock controlled *E4bp4* did not exert a rhythmic pattern in the heart of 19-day-old embryos but the distinct daily rhythm was observed in 4-day-old chicks with peak levels during the active phase of the photoperiod. Plasma glucose levels also exerted the daily rhythm on day 4 of postnatal life. To conclude, unlike the circadian oscillator in the pineal gland, peripheral clocks in the heart show the fully developed pattern of clock mechanisms after hatching. It is not known whether the peripheral pacemaker in the embryonic heart is not fully developed or transmission of signals and/or pathways from the master clock is not functioning during embryonic period.

Keywords: circadian; Per2; E4bp4; glucose; development; oscillator

Introduction

The avian circadian system regulates daily rhythms in metabolism and behavior. Its central part consists of the pineal gland, suprachiasmatic nuclei (SCN) and eyes (Cassone and Menaker, 1984). Expression of *cPer2* begins before day 15 of embryonic development in chicken SCN and on day 18 expression of *cPer2* is significantly higher during the light-time in comparison with dark-time under LD conditions (Okabayashi *et al.*, 2003). Rhythmic melatonin synthesis in the pineal gland begins several days before the end of the embryonic period and is entrainable by light and temperature cycles in precocial chicken (Zeman *et al.*, 1999; Zeman *et al.*, 2004). Expression of *cPer2* (Okabayashi *et al.*, 2003) and N-acetyltransferase (Herichova *et al.*, 2001) in the chicken pineal gland also show rhythmic

*Corresponding author
E-mail: herichova@fns.uniba.sk

pattern at this stage of development. In altricial European starlings the circadian clock in the pineal gland is operating immediately after hatch (Gwinner *et al.*, 1997). These finding strongly indicates that the avian circadian system develops prenataly and by the time of hatching is functional.

Clock mechanisms driving oscillations in mammals and birds are based on autoregulatory feed- back loops consisting of clock genes and transcriptional factors and possess a lot of common features (Okano and Fukada, 2003; Shearman *et al.*, 2000). Machinery generating oscillations in the brain were found also in the avian heart (Chong *et al.*, 2003). Since it is believed that peripheral clocks can modulate physiological functions we decided to study expression of clock genes in the heart during ontogeny and compare daily profiles of *cPer2* and *E4bp4* expression during late stage of embryonic development and early stage of postembryonic development in chicken. To monitor metabolic state that can strongly contribute to synchronisation of peripheral oscillators (Damiola *et al.*, 2000) we monitored plasma glucose concentrations.

Material and methods

Hatching eggs of broiler breeders were incubated in a forced draught incubator (BIOS Midi, Sedlcany, Czech Republic) at a temperature of 37.3±0.3 °C and relative humidity 55-65% and were turned automatically every two hours. Lighting was provided by a 18 Watt cool white fluorescent tube (Osram, Lumilux combi, Germany) that produced illumination in the range of 40-80 lx at the level of the eggs. During incubation eggs were exposed to the light (L) : dark (D) cycle 12:12 [light on 6:00h (Zeitgeber Time, ZT 0) and light off on 18:00h (ZT 12)]. After hatching chicks were kept in a brooder room with the same LD cycle as was installed in the incubator. The ambient temperature in the room was kept at 37°C for the first 2 days and 34°C for the next 2 days. Food and water were available *ad libitum*. During sampling in the dark period, red light with low intensity was used. Experimental protocols were approved by the Ethical Committee for the Care and Use of Laboratory Animals at the Comenius University Bratislava.

Experiment 1: The aim of the first experiment was to describe the 24-hr profile of *cPer2* and *E4bp4* expression in the heart of 19-day-old chick embryos and 4-day-old chicks. Sampling occurred in four hour intervals beginning at ZT 2. After decapitation samples of the heart tissues were immediately frozen in liquid nitrogen and stored under –80 °C until RNA extraction.

Total RNA was isolated with use of Tri reagent® (MRC, USA). The RNA integrity and contamination with DNA was examined by agarose gel electrophoresis. Preparations were without any visible DNA contamination. All RNA samples to be compared were reverse transcribed into cDNA at the same time. The first-strand cDNA synthesis was carried out with use of ImProm-II™ Reverse Transcription System and oligo(dT)15 primers (0.5 µg per reaction) according to the manufacturer instruction (Promega, USA). Denatured total RNA (1 µg) was used as template in a 20-µl cDNA synthesis reaction.

Aliquots of the cDNA (0.5-2 µl of RT product) were analysed for gene expression with the appropriate primers in 20-µl real time PCR reactions. The primers for the PCR analysis were as follows: for *s17* (X07257) forward primer 5'-ACACCCGTCTGGGCAACGAC-3', reverse primer 5'-CCCGCTGGATGCGCTTCATC-3' (Dragon et al., 2002); for *cPer2* (AF246956) forward primer 5'- ACCTAAAGGAAGGCCTGGTG-3', reverse primer 5'-CGCTGAGTAGCTTGCTTGTG-3'; for *E4bp4* (AF335427) forward primer 5'-

TGCCACACAGAAATTGTCATC-3', reverse primer 5'- CAACTCCAGTTTTGCAACCA-3';. Quantification of cDNA was performed with use QuantiTect SYBR Green PCR Kit (QIAGEN, Germany) and ABI PRISM® 7900HT Sequence Detection System (Applied Biosystems, USA). Real time PCR conditions were: 95°C, 15 min followed by 50 cycles of 94°C, 15 s; 47.5 °C, 30 s and 72°C, 30 s. Specificity and identity of the PCR product was validated by melting curve analysis and gel electrophoresis.

Experiment 2: The aim of the second experiment was to describe the 24-hr profile of glucose concentration in 19-day-old chick embryos and 4-day old chicks. Samples were taken in four hour intervals beginning at ZT 14. After decapitation trunk blood was collected in heparinised tubes and centrifuged under refrigeration at 2000g for 10 min. Plasma samples were collected and stored at -18°C until assay. Concentration of glucose in plasma was measured by standard kit Glucosa god 250 (PLIVA Lachema, CR) according manufacturer instructions.

Statistics: Daily profiles were fitted into a cosinor curve with 24 hr period and when experimental data significantly matched the cosinor curve its parameters were calculated with 95% confidence limits: mesor (the time series mean), amplitude (one-half the peak–trough difference expressed herein relative to the mesor) and acrophase (peak time referenced to the time of lights on in the animal facility). Since we determined gene expression in relative units only acrophase is stated for rhythmic pattern of *cPer2* and *E4bp4* expression. Goodness of fit (R-value) of the approximated sinusoidal curve was estimated by ANOVA (Nelsom *et al.*, 1979; Klemfus and Clopton; 1993).

Results and discussion

We did not find a significant rhythmic pattern in *cPer2* and *E4bp4* expression in the heart of 19-day-old chick embryos (Fig. 1A and 2A). Plasma glucose concentration did not show rhythmic changes at this stage of development either, and regression analysis confirmed a significant increase in glucose levels (Fig. 3A). This observation is in accordance with published data (Lu *et al.*, 2004).

Since expression of clock genes was not rhythmic in the embryonic heart we can conclude that functioning of the circadian system starts sequentially in birds. In 19-day-old embryos the central part of the circadian system is capable to drive melatonin rhythm but peripheral clocks in the heart are not working. It was shown that a daily pattern in systolic and diastolic blood pressure is not detectable at this age (Tazawa *et al.*, 2002) but appears at the time of hatching with no regard if chicken is actually hatched (Moriya *et al.*, 2000), indicating work of autonomous circadian clock.

Early after hatching expression of both genes exerted the distinct daily rhythm (Fig. 1B and 2B). Peak levels in *cPer2* expression appeared 3-4 hrs after light on and expression of *E4bp4* achieved highest levels 7 hrs after light on. Phase relationship between *cPer2* and *E4bp4* expression is similar to the one in the pineal gland (Doi *et al.*, 2004). Expression of *cPer2* is nearly in antiphase to expression *cBmal1* and *cMOP4* in the heart (Chong *et al.*, 2003). Similarly in rats (Young *et al.*, 2001), expression of *cPer2* in the heart peaks during the active phase. Rhythmic pattern of *cPer2* in the heart is therefore inversive in chicken and rat, but because of lack of information from diurnal mammals and/or nocturnal birds it is not possible to decide whether it is due to interspecies difference or is linked to activity of the animals.

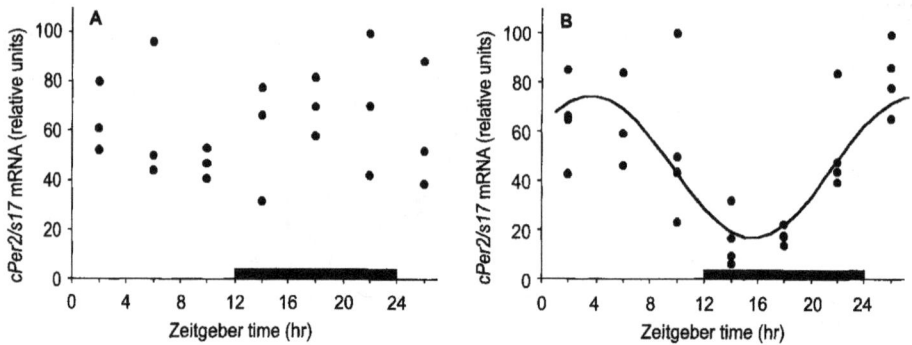

Figure 1. Daily expression of *cPer2* in the heart of 19-day-old embryos (A) and 4-day-old (B) chicks synchronized to LD cycle 12L:12D. The time of sampling is expressed in Zeitgeber Time (ZT), the onset of light-time is considered as ZT 0. Black bar on the bottom of the graph corresponds to dark-time. Results are expressed as individual measurements (n=3-4 per group). Black solid line displays the best fitted cosinor curve for postnatal *cPer2* expression (P<0.05; R=0.758).

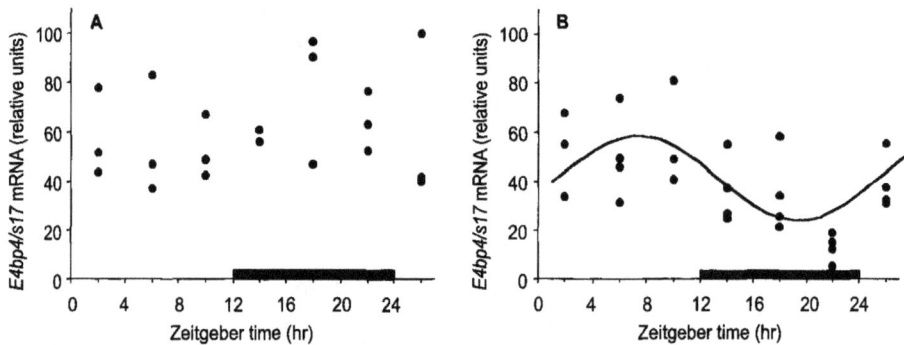

Figure 2. Daily expression of *E4bp4* in the heart of 19-day-old embryos (A) and 4-day-old (B) chicks synchronized to LD cycle 12L:12D. The time of sampling is expressed in Zeitgeber Time (ZT), the onset of light-time is considered as ZT 0. Black bar on the bottom of the graph corresponds to dark-time. Results are expressed as individual measurements (n=3-4 per group). Black solid line displays the best fitted cosinor curve for postnatal *E4bp4* expression (P<0.05; R=0.522).

Plasma glucose levels showed a daily rhythm with average of 12.32 mmol/l, amplitude 1.56 mmol/l and peak level 5.5 hrs after light on (Fig. 3B) which resembles the daily pattern of plasma glucose level in Japanese quail (Herichova *et al.*, 2004). Important role of interme-diatory metabolism in synchronisation of peripheral oscillators is proved in mammals (Damiola *et al.*, 2000) and very likely plays a crucial role also in birds.

Arhythmic daily pattern of *cPer2* and *E4bp4* in the chicken heart on day 19 of development can imply that autonomous oscillator in the heart and/or pathways important for its synchronisation are not developed yet. Plasma glucose did not exert the daily rhythm and many other metabolites also show an increasing trend in plasma concentrations during embryonic development because of activation of metabolism (Lu *et al.*, 2004). This can cause desynchronisation of peripheral oscillators or simply not initiate the functioning of oscillators. Another explanation can be that a crucial environmental synchronizing factor was missing.

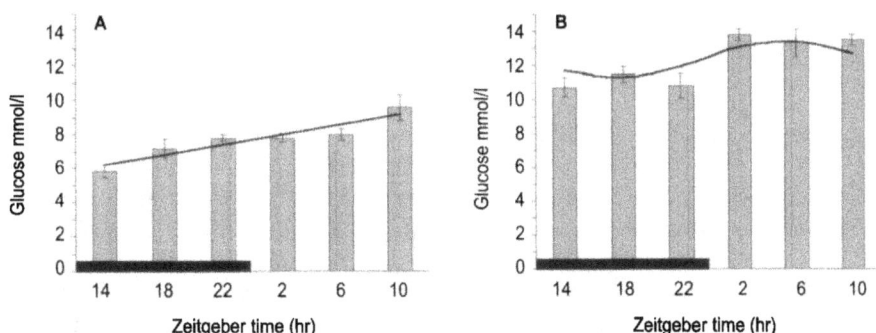

Figure 3. Daily rhythm of plasma glucose in 19-day-old embryos (A) and 4-day-old (B) chicks synchronized to LD cycle 12L:12D. The time of sampling is expressed in Zeitgeber Time (ZT), the onset of light-time is considered as ZT 0. Black bar on the bottom of the graph corresponds to dark-time. Results are expressed as mean ± S.E.M (n=4-6). Black solid line displays the best fitted linear regression for embryonic plasma glucose levels (P<0.05; R=0.930) and the best fitted cosinor curve for postembryonic plasma glucose levels (P<0.05; R=0.696).

It is possible that at this age the temperature cycle would be a more potent synchronising factor for the developing embryo than the light-dark cycle. Under natural conditions embryos are exposed to temperature cycles as the hen is leaving the nest because of feeding. In this way peripheral oscillators, especially those functionally linked to breathing and blood supply, can be initiated and synchronised.

Our study demonstrated for the first time expression of clock gene *cPer2* and *E4bp4* during prenatal life in the heart of precocial chicken. Our results indicate that avian circadian system begins to work successively and rhythmic expression of *cPer2* and *E4bp4* in the heart develops after hatch. Ways how central and peripheral components of the circadian system are interconnected and how their connection contributes to functional state of peripheral oscillators remain to be elucidate.

Acknowledgement

This study was supported by Science and Technology Assistance Agency under the contract No. APVT-20-022704.

References

CASSONE, V.M. and MENAKER, M. (1984) Is the avian circadian system a neuroendocrine loop? *The Journal of Experimental Zoology* **232**: 539-549.

CHONG, N.W., CHAURASIA, S.S., HAQUE, R., KLEIN, D.C. and IUVONE, P.M. (2003) Temporal-spatial characterization of chicken clock genes: circadian expression in retina, pineal gland, and peripheral tissues. *Journal of Neurochemistry* **85**: 851-860.

DAMIOLA, F., LE MINH, N., PREITNER, N., KORNMANN, B., FLEURY-OLELA, F. and SCHIBLER, U. (2000) Restricted feeding uncouples circadian oscillator in peripheral tissues from the central pacemaker in the suprachiasmatic nucleus. *Genes & Development* **14**: 2950-2961.

DOI, M., OKANO, T., YUJNOVSKY, I., SASSONE-CORSI, P. and FUKADA, Y. (2004) Negative control of circadian clock regulator E4BP4 by casein kinase I e-mediated phosphorylation. *Current Biology* **14**: 975-980.

DRAGON, S., OFFENHA, N. and BAUMANN, R. (2002) cAMP and *in vivo* hypoxia induce *tob*, *ifr1*, and *fos* expression in erythroid cells of the chick embryo. *American Journal of Physiology. Regulatory, Integrative and Comparative Physiology* **282**: R1219–R1226.

GWINNER, E., ZEMAN, M. and KLAASSEN, M. (1997) Synchronization by low-amplitude light-dark cycles of 24-hour pineal and plasma melatonin rhythms of hatchling European starlings (*Sturnus vulgaris*). *Journal of Pineal Research* **23**: 176-181.

KLEMFUSS, H. and CLOPTON, P.L. (1993) Seeking tau: a comparison of six methods. J. *Journal of Interdisciplinary Cycle Research* **24**: 1-16.

HERICHOVA, I., ZEMAN, M., MACKOVA, M. and GRIAC, P. (2001) Rhythms of the pineal N-acetyltransferase mRNA and melatonin concentrations during embryonic and post-embryonic development in chicken. *Neuroscience letters* **298**: 123-126.

HERICHOVA, I., ZEMAN, M., JURANI, M. and LAMOSOVA, D. (2004) Daily rhythms of melatonin and selected biochemical parameters in plasma of Japanese quail. *Avian and Poultry Biology Reviews* **15**: 205-210.

LU, J., MCMURTRY, J.P. and COON, C.N. (2004) Developmental changes of plasma insulin, glucagons, Igf-I, Igf-Ii, thyroid hormones, and glucose concentration in chick embryo. *Poultry Science.* **83**(SUPPL.1)**:** 192.

MORIYA, K., PEARSON, J.T., BURGGREN, W.W., AR, A. and TAZAWA, H. (2000) Continuous measurements of instantaneous heart rate and its fluctuations before and after hatching in chickens. *The Journal of experimental biology* **203**: 895-903.

NELSON, W., TONG, Y.L., LEE, J.K. and HALBERG, F. (1979) Methods for cosinor-rhythmometry. *Chronobiologia* **6**: 305-323.

OKABAYASHI, N., YASUO, S., WATANABE, M., NAMIKAWA, T., EBIHARA, S. and YOSHIMURA, T. (2003) Ontogeny of circadian clock gene expression in the pineal and the suprachiasmatic nucleus of chick embryo. *Brain Research* **990**: 231-234.

OKANO, T. and FUKADA, Y. (2003) Chicktacking pineal clock. *Journal of Biochemistry (Tokyo)* **134**: 791-779.

SAKAMOTO, K., OISHI, K., NAGASE, T., MIYAZAKI, K. and ISHIDA, N. (2002) Circadian expression of clock genes during ontogeny in the rat heart. *Neuroreport* **13**: 1239-1242.

SHEARMAN, L.P., SRIRAM, S., WEAVER, D.R., MAYWOOD, E.S., CHAVES, I., ZHENG, B., KUME, K., LEE, C.C., VAN DER HORST, G.T., HASTINGS, M.H. and REPPERT, S.M. (2000) Interacting molecular loops in the mammalian circadian clock. *Science* **288**: 1013-1019.

TAZAWA, H., AKIYAMA, R. and MORIYA, K. (2002) Development of cardiac rhythms in birds. *Comparative Biochemistry and Physiology. Part A, Molecular & Integrative Physiology* **132**: 675-689.

YOUNG, M.E., RAZEGHI, P. and TAEGTMEYER, H. (2001) Clock genes in the heart: characterization and attenuation with hypertrophy. *Circulation Research* **88**: 1142-1150.

ZEMAN, M., PAVLIK, P., LAMOSOVA, D., HERICHOVA, I. and GWINNER, E. (2004) Entrainment of rhythmic melatonin production by light and temperature in the chick embryo. *Avian and Poultry Biology Reviews* **15**:197-204.

ZEMAN, M., GWINNER, E., HERICHOVA, I., LAMOSOVA, D. and KOSTAL, L. (1999) Perinatal development of circadian melatonin production in domestic chicks. *Journal of Pineal Research* **26**: 28-34.

Enhancement of meat production by thermal manipulations during embryogenesis of broilers

ORNA HALEVY[1*], MAYAN LAVI[1], SHLOMO YAHAV[2]

[1]Dept. of Animal Sciences, Faculty of Agricultural, Food and Environmental Quality Sciences, The Hebrew University of Jerusalem, Rehovot 76100, Israel; [2]ARO, The Volcani Center, Institute of Animal Sciences, P.O. Box 6 Bet Dagan 50250, Israel

This study was aimed at elucidating the effect of thermal manipulation (TM) at late-term broiler embryos on muscle growth and development during the early growth phase posthatch. We hypothesized that TM causes an increase in satellite cell proliferation, necessary for further muscle hypertrophy. The TM was conducted on embryonic days E16 to E18 at 38.5°C and 65% RH for 3 hours daily. Muscle percentage of body weight was higher in the TM chicks than in controls on day 9 of age. An increase was observed in satellite cell proliferation in culture and *in vivo* in response to TM on days 6 and 9 posthatch, to levels that were significantly higher than those of control chicks. This was accompanied by a marked induction of the muscle regulatory factor myogenin as well as of insulin-like growth factor-I (IGF-I) levels in the breast muscle. The kinetics of myogenin and Pax7 expression in muscle cells was delayed in response to the TM and peaked later than in controls. These data suggest that under the conditions of this study, TM delays satellite cell differentiation and allows these cells to remain in the cell cycle longer. This results in more cell proliferation and subsequently, in enhanced muscle growth and meat production.

Introduction

Development and histogenesis of skeletal muscle proceeds from early embryogenesis through adulthood. Embryonic muscle-precursor cells undergo myogenic determination and give rise to myoblasts (Ordhal *et al.*, 2000). Myoblasts proliferate, withdraw from the cell cycle, differentiate, and eventually fuse into multinucleated fibers. This myogenic program involves the expression of a cascade of muscle-specific genes that is regulated by members of the myogenic basic helix-loop-helix (bHLH) transcription factors, including the MyoD and myocyte-enhancer factor-2 (MEF2) families (Naya and Olson, 1999).

Previous studies have indicated that the developing muscle consists of multiple myogenic populations (Cossu and Molinaro, 1987; Stockdale, 1992). In the chick, embryonic myoblasts are most abundant on E5, whereas fetal myoblasts are most abundant between E8 and E12 (Stockdale, 1992). Individual myofibers become encased by a basement membrane during late embryogenesis (E15 onwards), and it is at this stage that it becomes possible to distinguish satellite cells by their morphology and location (Hartley *et al.*, 1992; Yablonka-Reuveni, 1995). Satellite cells (or adult myoblasts), first identified by Mauro (1961), are the

*Corresponding author
E-mail: halevyo@agri.huji.ac.il

primary source of myogenic precursors in the postnatal muscle (Campion, 1984; Schultz and McCormik, 1994). These mononucleated cells lie under the basal lamina of the myofiber and are uniformly distributed throughout the length of the muscle (Campion, 1984). Skeletal muscle nuclei consist of a high percentage of proliferating satellite cells at hatch. However, in adults, the number of satellite cells decreases to less than 5% of total myofiber nuclei and they become largely quiescent (Hawke and Garry, 2001). Satellite cells re-enter the cell cycle in response to various muscular stresses and undergo proliferation followed by withdrawal from the cell cycle and fusion into existing or newly formed fibers (Bischoff, 1994; Grounds, 1998). Upon satellite cell activation, muscle-specific transcription factors are expressed in a sequential pattern with Myf5 and MyoD being expressed in the proliferating progeny followed by myogenin expression as the cells enter differentiation (Cooper *et al.*, 1999; Cornelison and Wold, 1997; Yablonka-Reuveni and Paterson, 2001). In addition, the expression of Pax7, the paired-box-containing transcription factor, is high during myoblast proliferation and decreases in association with differentiation in mouse myogenic cultures (Seale *et al.*, 2000). Pax7 is an early marker of myogenesis during adult muscle growth and its expression is maintained by satellite cells in chicken (Halevy *et al.*, 2004) and mouse adult muscle (Zammit *et al.*, 2004).

Several growth factors, including members of the fibroblast growth factor (FGF) family, (Florini *et al.*, 1996; Olwin *et al.*, 1994), platelet-derived growth factor (PDGF; Yablonka-Reuveni and Seifert, 1993) and hepatocyte growth factor (HGF; Allen *et al.*, 1995; Gal-Levi *et al.*, 1998; Leshem *et al.*, 2000, 2002) are able to stimulate satellite cell proliferation and inhibit differentiation. Insulin-like growth factor I (IGF-I) has been shown to stimulate proliferation as well as differentiation of satellite cells and increase myofiber hypertrophy (Adams and McCue, 1998; Coleman *et al.*, 1995; Florini *et al.*, 1996; Paul and Rosenthal, 2002).

Recent studies have demonstrated that mild heat stress (thermal conditioning, TC) for 24 h at 3 days of age induces compensatory growth, leading to improved performance and muscle growth in broilers (Yahav and Plavnik, 1999). This is due to enhanced proliferation and differentiation of satellite cells immediately after the TC period, which is affected by elevated levels of locally, secreted IGF-I (Halevy *et al.*, 2001).

This study analyzes the effects of temperature manipulations (TM) during broiler embryogenesis on muscle development at the early growth phase in posthatch broilers. TM during embryogenesis is uniform and perhaps more efficient at inducing alterations in fetal myoblast differentiation and adult myoblast proliferation. The E16 to E18 period was chosen because this is when fetal myoblasts undergo massive differentiation and adult myoblasts proliferate (Hartley *et al.*, 1992; Stockdale, 1992).

Materials and methods

EXPERIMENTAL PROCEDURE

Fertile Cobb strain broiler eggs (*n* = 120) were purchased from a local hatchery (Braun, Israel). The eggs were arranged in homological locations in two incubators. The incubators (Masalles, Spain, Type 65Hs) were identical and automatic. Incubation conditions from day 0 to day 21 were: 37.8°C and 56% RH for the control group. Thermal treatment of the eggs from E16 to E18 involved an increase in temperature to 38.5°C and in RH to 65% for 3 h (09:00-12:00) on each of those days. Immediately after the thermal treatments were terminated, incubation conditions were restored to regular levels. Eggs in both incubators were turned through 270° every hour. Data loggers (Microlog from Fourier Systems, UIL Ltd.) were placed in each incubator to monitor the conditions every 30 min. On E10, infertile and undeveloped eggs were removed after candling. On E19, the eggs were transferred to hatching trays located in each incubator. Upon hatching, each chick was weighed, sexed and wing-banded. Males were transferred to temperature-controlled brooders with free access to commercial diet and water.

CELL CULTURES

Skeletal muscle satellite cells were cultured from the pectoral muscle of male chicks at various ages as previously described (Halevy and Lerman, 1993; Halevy *et al.*, 2000). Cells were counted using a hemocytometer, and plated at $5x10^4$ cells/cm^2 in Dulbecco's Modified Eagle's Medium (DMEM) supplemented with 10% (v/v) horse serum (HS) on gelatin-coated dishes, and maintained at 37°C in a humidified atmosphere containing 95% air and 5% CO_2. On all days, cells were prepared under exactly the same conditions from 6 g of breast muscle that had been pooled from the experimental birds.

WESTERN BLOT ANALYSIS

Western blot analysis was performed as described in Leshem *et al.* (2002). In brief, muscle tissue was homogenized with a Kinematica homogenizer (Lucerne, Switzerland) for 30 s on ice in lysis buffer. Protein samples were sonicated and normalized to protein content (BCA kit, Pierce, Rockford, IL). Equal amounts of protein were subjected to direct immunoblotting for protein expression levels. Densitometric analysis was performed on bands using the Image Pro Plus 3.0 software. The following primary antibodies were used: a rabbit polyclonal antibody against chicken myogenin (a kind gift from Bruce Paterson, NIH, Bethesda, MD) and a mouse monoclonal antibody against IGF-I (Upstate Biotechnology, Lake Placid, NY).

INDIRECT IMMUNOFLUORESCENCE

Freshly prepared cells were grown for 1 day in 10% (v/v) HS-DMEM, after which they were fixed and permeabilized in PBS with 2% (w/v) formaldehyde for 15 min (Leshem *et al.*, 2002), and stained with a monoclonal antibody against Pax7 (hybridoma supernatant, Developmental Studies Hybridoma Bank, University of Iowa, IA) or anti-myogenin. Immune complexes were detected using FITC green-conjugated donkey anti-mouse and rhodamine-conjugated donkey anti-rabbit IgG. Nuclei were visualized by 4',6-diamidino-2-phenylindole (DAPI) staining.

PCNA ANALYSIS

Breast-muscle samples were removed from the same longitudinal region and immediately fixed in fresh 4% paraformaldehyde in PBS (pH 7.6), dehydrated and embedded in paraffin. Sections (5 μm) were cut, placed on glass slides, deparaffinized and rehydrated as previously described (Halevy *et al.*, 2004). Muscle sections were immunostained with an antibody against proliferating cell nuclear antigen (PCNA, a marker for dividing cells), using a commercial kit from Zymed (San Francisco, CA) followed by counterstaining with haematoxylin as previously described (Halevy et al., 2001). Control slides, where the primary antibody was omitted, were processed in parallel. Three chicks were analyzed per each group; five sections were studied per each chick, monitoring five random fields per each section. Analysis of positive cells was performed based on digitized images as previously described (Halevy et al., 2001).

STATISTICAL ANALYSIS

Data were subjected to analysis of variance (one-way ANOVA) and to Student's t-test, by means of the JMP® software (SAS Institute, 2000).

Results

Breast-muscle percentage of BW was equal for the first 6 days of age in both the TM and control groups, but was significantly higher in the TM group than in controls at 9 days of age (Fig. 1). The higher breast-muscle percentage suggested a potential effect of TM on muscle-cell proliferation. Muscle cells were prepared from breast muscle on E18 and on various days posthatch, and cell number per gram of muscle was determined. The number of muscle cells per gram of muscle did not vary between the groups on E18 (data not shown) or at hatch (Fig. 2A and B). Muscle-cell number per gram of muscle was lower in the TM group than in the control group until 3 days of age. However, at 6 and 9 days of age, the number of cells was significantly higher in the TM versus control group (Fig. 2B). Indeed, analysis of the expression levels of PCNA, a marker for cell proliferation, in breast muscle sections derived from the experimental chicks on day 9, revealed that the fold number of PCNA-expressing cells was 1.35-fold higher in the TM than in the control group (1.35 ± 0.037 and 1.00 ± 0.02, respectively; $P < 0.05$).

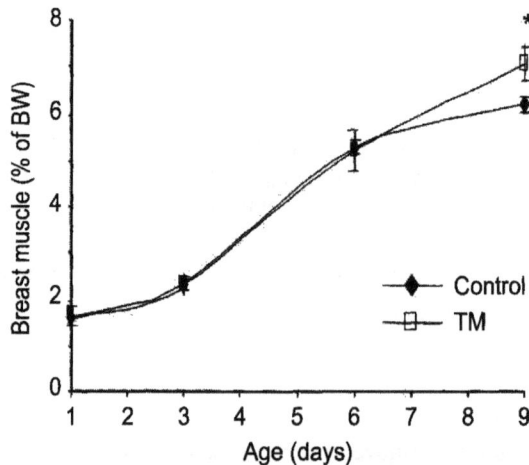

Figure 1. Breast muscle as percentage of body weight (BW) of chicks that were either thermally manipulated on embryonic days E16-E18 (TM) or served as untreated controls. Results are means ± SD; $n = 6$. *$P < 0.05$ vs. control at the same age.

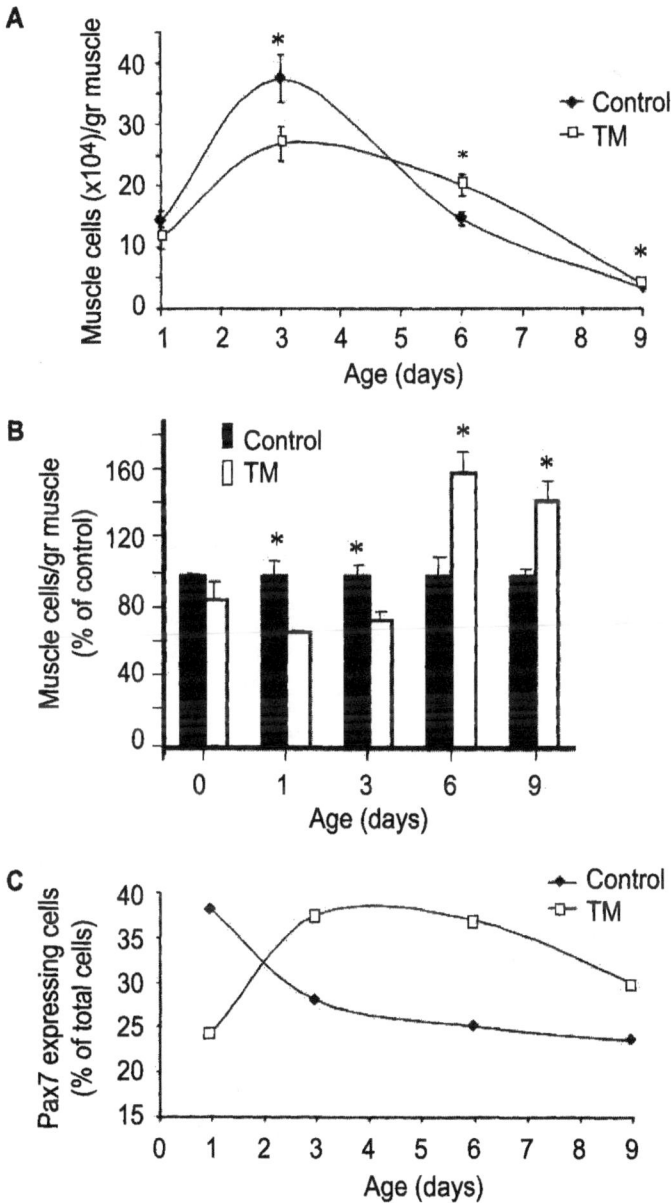

Figure 2. Muscle cells (satellite cells) per gram of breast muscle as absolute numbers (A) or as percentage of control (B) in TM and control chicks. Muscles were removed from the experimental chicks, pooled within each group and cells were prepared under similar conditions and counted. Results are means ±SD of three independent experiments. *$P < 0.05$ vs. control at the same age. (C) Pax7 expression in muscle cells in response to TM. Muscle cells were prepared as described in (A), and incubated for 17 h, after which they were fixed with 2% paraformaldehyde, and immunostained for Pax7. Nuclei were stained with DAPI. Results are presented as percent myogenin-expressing cells of the total DAPI-expressing cells and are the average of two independent experiments.

Indirect immunofluorescence assay for Pax7, an early marker for satellite cell proliferation (Halevy *et al.*, 2004), revealed that in muscle cell culture derived from the control
group, the number of Pax7-expressing cells was highest on day 1 posthatch after which it
declined, in agreement with previous reports (Halevy *et al.*, 2004; Fig. 2C). However, in
muscle cells derived from the TM group, the number of Pax7-expressing cells was the
highest between days 3 and 6 of age (Fig. 2C).

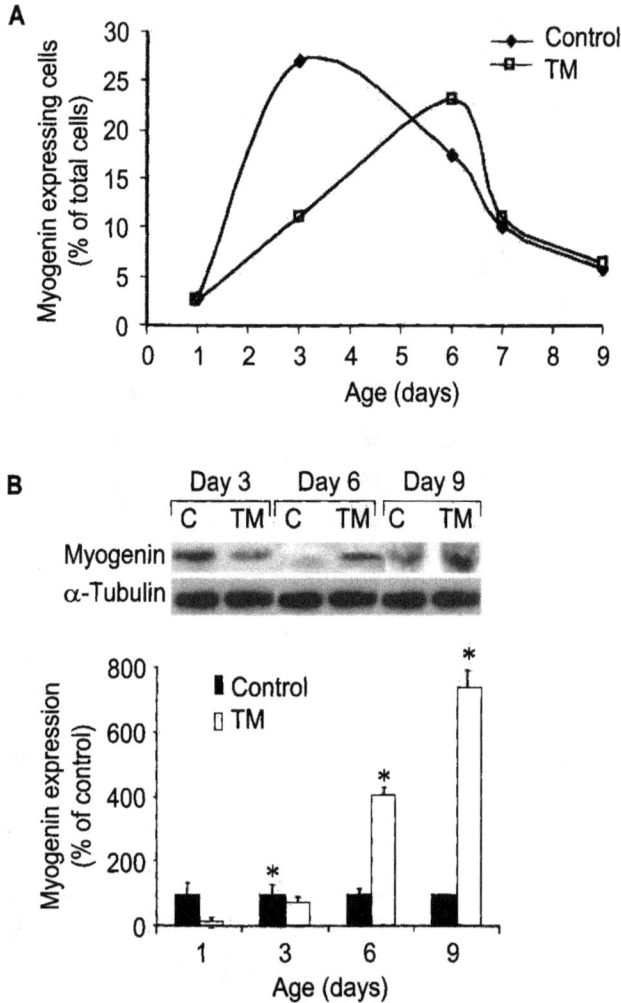

Figure 3. Myogenin expression in response to thermal manipulation during late-term embryogenesis.
(A) Muscle cells were prepared and immunostained for myogenin as described in Fig. 2. Results are
presented as percent myogenin-expressing cells of the total DAPI-expressing cells and are the average
of two independent experiments. (B) Upper panel: A representative western blot analysis for myogenin
expression in muscle tissues on various days posthatch. Equal quantities of protein were loaded as
evidenced by the α-tubulin bands. Lower panel: Densitometric analysis of myogenin expression relative to loaded protein levels in breast muscle samples derived from control and TM chicks at various
days of age. Results are means ± SE and are presented as percentage of control; $n = 4$. *$P < 0.05$ vs.
control at the same age.

In a similar assay, the percentage of myogenin-positive cells, a specific marker for muscle cell differentiation, peaked at 3 days of age in muscle cells derived from the control group, whereas it peaked on day 6 in the TM group (Fig. 3A). Similar results were obtained in whole breast muscle extracts from the experimental chicks (Fig. 3B). Myogenin protein levels were lower in muscles derived from the TM chicks than in those from controls on days 1 and 3. However, they were significantly higher on days 6 and 9, reaching four- and sevenfold differences, respectively (Fig. 3B, lower panel).

Densitometric analysis of the breast muscle samples derived from chicks at various days of age revealed up to fourfold higher IGF-I protein expression in muscles derived from the TM vs. control group, on days 3 and 9 of age (Fig. 4). The difference in IGF-I levels between the groups was less pronounced on day 6, but was nevertheless significantly higher in the TM group.

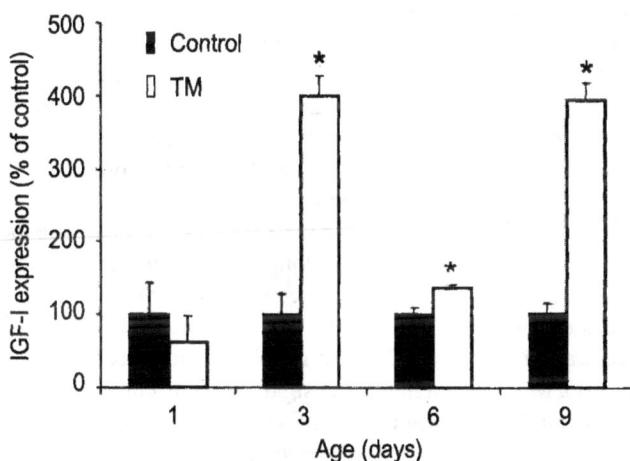

Figure 4. Densitometric analysis of IGF-I expression relative to loaded protein levels in breast muscle samples derived from control and TM chicks at various days of age. Results are means ± SE, and presented as percentage of control; $n = 4$. *$P < 0.05$.

Discussion

The results of this study suggest that short periods of heat exposure (38.5°C) during late-term embryogenesis affect muscle cell proliferation and differentiation as well as muscle growth on the early phase of growth in posthatch broilers.

The induction in cell proliferation in response to the TM was evident in adult myoblasts (satellite cells) posthatch but not in myogenic cells prior to, or at, hatching day. In fact, higher numbers of muscle cells as well as a higher number of PCNA-expressing cells were found in the TM group than in controls not before days 6 and 9 posthatch, suggesting an on-going proliferative activity of satellite cells on late rather than on very early days posthatch. Myogenin has been reported to be an early marker of muscle-cell differentiation, as it marks the transition of proliferating cells to differentiating ones (Andres and Walsh, 1996; Yablonka-Reuveni and Paterson, 2001). Our finding that muscle myogenin levels were significantly higher in the TM group relative to controls on these later days suggests that alongside a still-proliferating cell population, a large cell population is about to differentiate and fuse into myofibers. Increased muscle regulatory factor levels have been attributed to satellite cells and not myofibers at an early age (Adams *et al.,* 1999; Yablonka-Reuveni, 1995).

The enhanced proliferation and differentiation of satellite cells in the TM group could be attributed to the higher muscle IGF-I levels found on days 3 and 6, suggesting a role for IGF-I in the modulation of satellite cell proliferation and differentiation following thermal manipulations. We believe that the higher IGF-I levels found in the TM muscle represent the locally secreted growth factor (i.e. in the muscle), as was found in the case of exposing 3-day-old chicks to mild heat stress (Halevy *et al.,* 2001) and in that of induced hypertrophy by localized IGF-I infusion (Adams and McCue, 1998). The second rise in IGF-I in the TM group on day 9 may in part explain the higher hypertrophy in this group, as IGF-I has been reported to play a major role in muscle hypertrophy (Coleman *et al.,* 1995 Adams *et al.,* 1999). Moreover, muscle IGF-I levels strongly correlate with muscle mass (Adams *et al.,* 2000; Paul and Rosenthal, 2002).

At this stage, it is unclear whether regulation of IGF-I in the muscle is directly dependent on TM or indirectly affected by other factors. One of these factors could be the thyroid hormones. Recently, it has been reported that thyroid deficiency decreases muscle IGF-I levels, specifically perturbing muscle growth in rats (Adams *et al.,* 2000). The levels of thyroid hormones were found to be significantly lower at hatch in chicks that had undergone TM compared to untreated chicks on E16 through E18, under conditions similar to those in this study (Yahav *et al.,* 2004). This may explain, at least in part, the relatively lower IGF-I levels and lower cell proliferation on day 1 in the TM chicks compared to the untreated ones. Conversely, higher IGF-I expression levels along with compensatory muscle growth were observed in the TM chicks from 3 days of age onward. Compensatory growth has been observed in chicks exposed to TC at 3 days posthatch and this coincided with increased plasma T3 concentration (Yahav and McMurtry, 2001), and muscle IGF-I expression levels (Halevy *et al.,* 2001). Moreover, a linear and significant correlation was previously demonstrated between plasma T3 and feed intake and weight gain in the domestic fowl (Yahav *et al.,* 1998; Yahav, 2000), suggesting a similar pattern of T3 levels in the TM chicks. In view of these findings, it may well be that IGF-I's effects on muscle growth in response to TM are at least in part mediated by thyroid hormones.

Previously, we have shown that TC in Cobb broilers at 3 days of age immediately enhances proliferation and accelerates differentiation of satellite cells (Halevy *et al.,* 2001). In contrast, here there was no immediate proliferative response after the heat manipulation (i.e., on E18 and at hatch). Rather, response to TM was delayed as observed for IGF-I expression levels in muscle and Pax7 and myogenin expression levels in muscle cells in the TM group versus the control group. Moreover, the decline in cell number was more moderate in the TM group than in the control (Fig. 2A), resulting in higher cell numbers on day 9. Normally in broilers, satellite cell numbers peak at 3 to 4 days of age and then rapidly decline (Halevy *et al.,* 2000, 2001, 2004). Together, these findings suggest that at least under this study's conditions, embryonic TM delays, rather than accelerates the kinetics of satellite cell myogenesis. However, we cannot rule out the possibility of an immediate proliferative response, as was found in a study during which TM was conducted at 39.5°C from E16 to E18, under a similar schedule (Piestun, Y., Halevy, O., Yahav, S; unpublished results). Collectively, the findings suggest that satellite cell myogenesis is highly affected by TM in the perinatal period. It is conceivable that combinations of temperature, timing and duration of the TM are critical for processes that underlie muscle growth regulation in broilers. However, the exact relationship between TM in the embryo and their effect on muscle growth needs to be further studied.

In summary, this study focused on alterations in muscle growth and development during the early growth phase post hatch in response to TM. We propose that TM during the late stage of incubation influence muscle growth in posthatch broilers. Under the conditions of this study, they delay satellite-cell differentiation and allow these cells to remain in the cell cycle longer. This results in more cells and subsequently, in enhanced muscle growth and meat production.

Acknowledgements

We are grateful to Bruce Paterson for providing the chicken myogenin antibody. We thank D. Shinder and R. Sasson-Rat for their assistance. O. Halevy is a holder of the Charles Charkowsky Chair in Poultry Science and Animal Hygiene. This work was supported in part by the Israeli Poultry Marketing Board.

References

ADAMS, G.R., and McCUE, S.A. (1998) Localized infusion of IGF-I results in skeletal muscle hypertrophy in rats. *Journal of Applied Physiology* **84**: 1716-1722.

ADAMS, G.R., HADDAD, F., and BALDWIN, K.M. (1999) Time course of changes in markers of myogenesis in overload rat skeletal muscles. *Journal of Applied Physiology* **85**: 1705-1712.

ADAMS, G.R., McCUE S.A., BODELL, B.W., ZENG, M., and BALDWIN, K.M. (2000) Effects of spaceflight and thyroid deficiency on hindlimb development. I. Muscle mass and IGF-I expression. *Journal of Applied Physiology* **88**: 894-903.

ALLEN, R.E., SHEENAN, S.M., TAYLOR, R.G., KENDALL, T.L., and RICE, G.M. (1995) Hepatocyte growth factor activates quiescent skeletal muscle satellite cells *in vitro. Journal of Cellular Physiology* **165**: 307-312.

ANDRES, V., and WALSH, K. (1996) Myogenin expression, cell cycle withdrawal, and phenotypic differentiation are temporally separable events that precede cell fusion upon myogenesis. *Journal of Cell Biology* **132**: 657-666.

BISCHOFF, R. (1994) The satellite cell and muscle regeneration. Pages 97-118 in: Engel, A.G., Franzini-Armstrong, C., editors. Myology. New York: McGraw-Hill.

CAMPION, D.R. (1984) The muscle satellite cell: a review. *International Review of Cytology* **87**: 225-251.

COOPER, R.N., TAJBAKHSH, S., MOULY, V., COSSU, G., BUCKINGHAM, M., and BUTLER-BROWNE, G.S. (1999) *In vivo* satellite cell activation via Myf5 and MyoD in regenerating mouse skeletal muscle. *Journal of Cell Science* **112**: 2895-2901.

COLEMAN, M.E., DeMAYO, F., YIN, K.C., LEE, H.M., GESKE, R., MONTGOMERY, C., and SCHWARTZ, R.J. (1995) Myogenic vector expression of insulin-like growth factor I stimulates muscle cell differentiation and myofiber hypertrophy in transgenic mice. *Journal of Biological Chemistry* **270**: 12109-12116,

CORNELISON, D.D.W., and WOLD, B. (1997) Single-cell analysis of regulatory gene expression in quiescent and activated mouse skeletal muscle satellite cells. *Developmental Biology* **191**: 270-283.

COSSU, G., and MOLINARO, M. (1987) Cell heterogeneity in the myogenic lineage. *Current Topics in Developmental Biology* **23**: 185-208.

FLORINI, J.R., EWTON, D.Z., and COOLICAN, S.A. (1996) Growth hormone and the insulin-like growth factor system in myogenesis. *Endocrine Reviews* **17**: 481-517.

GAL-LEVI, R., LESHEM, Y., AOKI, S., NAKAMURA, T., and HALEVY, O. (1998) Hepatocyte growth factor plays a dual role in regulating skeletal muscle satellite cell proliferation and differentiation. *Biochimica et Biophysica Acta* **1402**: 39-51.

GROUNDS, M.D. (1998) Age-associated changes in the responses of skeletal muscle cells to exercise and regeneration. *Annals of the New York Academy of Sciences* **854**: 78-91.

HALEVY, O., and LERMAN, O. (1993) Retinoic acid induces adult muscle cell differentiation mediated by the retinoic acid receptor-a. *Journal of Cellular Physiology* **154**: 566-572.

HALEVY, O., GEYRA, A., BARAK, M., UNI, Z., and SKLAN, D. (2000) Early posthatch starvation decreases satellite cell proliferation and skeletal muscle growth in chicks. *Journal of Nutrition* **130**: 858-864.

HALEVY, O., KRISPIN, A., LESHEM, Y., McMURTRY, J.P. and YAHAV, S. (2001) Early-age heat exposure affects skeletal muscle satellite cell proliferation and differentiation in chicks. *American Journal of Physiology* **281**: R302-R309.

HALEVY, O., PIESTUN, Y., ALLOUTH, M., ROSSER, B., RINKEVITCH, Y., RESHEF, R., ROZENBOIM, I., WLEKLINSKI, M., and YABLONKA-REUVENI, Z. (2004) The pattern of Pax-7 expression during myogenesis in the posthatch chicken establishes a model for satelite cell differentiation and renewal. *Developmental Dynamics* **231**: 489-502.

HARTLEY, R.S., BANDMAN, E., and YABLONKA-REUVENI, Z. (1992) Skeletal muscle satellite cells appear during late chicken embryogenesis. *Developmental Biology* **153**: 206-216.

HAWKE, T.J., and GARRY, D.J. (2001) Myogenic satellite cells: physiology to molecular biology. *Journal of Applied Biology* **91**: 534-551.

LESHEM, Y., SPICER, D.B., GAL-LEVI, R., and HALEVY, O. (2000) Hepatocyte growth factor (HGF) inhibits skeletal muscle cell differentiation: a role for the bHLH protein twist and the cdk inhibitor p27. *Journal of Cellular Physiology* **184**: 101-109.

LESHEM, Y., GITELMAN, I., PONZETTO, C., and HALEVY, O. (2002) Preferential binding of Grb2 or phosphatidylinositol 3-kinase to the Met receptor has opposite effects on HGF-induced myoblast proliferation. *Experimental Cell Research* **274**: 288-298.

MAURO, A. (1961) Satellite cells of skeletal muscle fibers. *Journal of Biophysics and Biochemistry Cytology* **9**: 493-495.

NAYA, F.S., and OLSON, E. (1999) MEF2: a transcriptional target for signaling pathways controlling skeletal muscle growth and differentiation. *Current Opinion in Cell Biology* **11**: 683-688.

OLWIN, B.B., HANNON, K., and KUDLA, A.J. (1994) Are fibroblast growth factors regulators of myogenesis *in vivo*? *Progress in Growth Factor Research* **5**: 145-158.

ORDAHL, C.P., WILLIAMS, B.A., and DENETCLAW, W. (2000) Determination and morphogenesis in myogenic progenitor cells: an experimental embryological approach. *Current Topics in Developmental Biology* **48**: 319-367.

PAUL, A.C., and ROSENTHAL, N. (2002) Different modes of hypertrophy in skeletal muscle fibers. *Journal of Cell Biology* **156**: 751-760.

SAS INSTITUTE (2000) JMP® Statistics and Graphics Guide, Version 4. Cary, NC: SAS Institute Inc.

SCHULTZ, E., and McCORMIK, K.M. (1994) Skeletal muscle satellite cells. *Reviews in Physiology and Biochemical Pharmacology* **123**: 213-257.

SEALE, P., SABOURIN, L.A., GIRGIS-GABARDO, A., MANSOURI, A., GRUSS, P., and RUDNICKI, M.A. (2000) Pax7 is required for the specification of myogenic satellite cells. *Cell* **102**: 777-786.

STOCKDALE, F.E. (1992) Myogenic cell lineages. *Developmental Biology* **154**: 284-298.

YABLONKA-REUVENI, Z. (1995) Myogenesis in the chicken: the onset of differentiation of adult myoblasts is influenced by tissue factors. *Basic Applied Myology (BAM)* **5**: 33-41.

YABLONKA-REUVENI, Z., and SEIFERT, R.A. (1993) Proliferation of chicken myoblasts is regulated by specific isoforms of platelet-derived growth factor: evidence for differences between myoblasts from mid and late stages of embryogenesis. *Developmental Biology* **156**: 307-318.

YABLONKA-REUVENI, Z., and PATERSON, B. (2001) MyoD and myogenin expression patterns in cultures of fetal and adult chicken myoblasts. *Journal of Histochemistry and Cytochemistry* **49**: 455-462.

YAHAV, S. (2000) Relative humidity at moderate temperatures, its effect on male broiler chickens and turkeys. *British Poultry Science* **41**: 94-100.

YAHAV, S., and PLAVNIK, I. (1999) The effect of an early age thermal conditioning and food restriction on performance and thermotolerance of male broiler chickens. *British Poultry Science* **40**: 120-126.

YAHAV, S., and McMURTRY, J. (2001) Thermotolerance acquisition in broiler chickens by temperature conditioning early in life—the effect of timing and ambient temperature. *Poultry Science* **80**: 1662-1666.

YAHAV, S., PLAVNIK, I., RUSAL, M., and HURWITZ, S. (1998) The response of turkeys to relative humidity at high ambient temperature. *British Poultry Science* 39: 340-345.

YAHAV, S., SASSON RATH, R., and SHINDER, D. (2004) The effect of thermal manipulations during embryogenesis of broiler chicks (*Gallus domesticus*) on hatchability, body weight and thermoregulation after hatch. *Journal of Thermal Biology* 29: 245-250.

ZAMMIT, P.S., GOLDING, J.P., NAGATA, Y., HUDON, V., PARTRIDGE, T.A., and BEAUCHAMP, J.R. (2004) Muscle satellite cells adopt divergent fates: a mechanism for self-renewal? *Journal of Cell Biology* **166**: 347-357.

The myogenic regulatory factors and the neurogenic factor nAChR are differently affected in fast and slow muscles by neuromuscular stimulation in developing avian embryos

G.M. MCENTEE, B.H. SIMBI, & N.C. STICKLAND

Department of Veterinary Basic Sciences, The Royal Veterinary College, Royal College Street, London, NW1 0TU

The pre-natal manipulation of environmental factors such as maternal nutrition, movement and temperature can affect the development of muscle by enhancing fibre and nuclear number. We investigated the mechanisms involved in the expression of genes associated with muscle development by enhancing neuromuscular stimulation using 4-aminopyridine applied *in ovo* from embryonic days 10-13. A fast (posterior *latissimus dorsi*) and slow (anterior *latissimus dorsi*) muscle from the control and treated groups were removed and snap-frozen. Embryos were sampled from ED15-ED20. The mRNA expression of MyoD, myogenin, myostatin, MRF4 and nAChR were analysed using RT-PCR. Data were normalised and analysis of variance applied. There was a down-regulation of nAChR in the anterior *latissimus dorsi* (ALD) of the treated compared to the control group, while there was an up-regulation in the expression of nAChR in the treated posterior *latissimus dorsi* (PLD) compared to the control group. There was an up-regulation of myostatin mRNA expression on ED18 and a suppression of myogenin mRNA on ED15 in the treated ALD group. There was more myostatin and less myogenin mRNA expressed in the stimulated compared to the control group of the PLD. These results suggested a delay in differentiation allowing for a possible increase in the pool of proliferating myoblasts and satellite cells. Therefore, increased neuromuscular stimulation may have a positive effect on secondary muscle fibre development, possibly by increasing myoblast proliferation resulting in the delay of myogenin expression through prolonged innervation during secondary myogenesis.

Key words: MRFs, Neuromuscular junction, embryonic muscle, stimulation, chick

Introduction

The pre-natal manipulation of environmental factors such as maternal nutrition (Bayol *et al.*, 2004) and temperature (Maltby *et al.*, 2004) can significantly affect the development of skeletal muscle by influencing fibre and nuclear number. Our previous results have indicated that neuromuscular-stimulated movement using 4-aminopyridine (4-AP) can significantly increase fibre number immediately pre-hatch at embryonic day 20 (see Heywood *et al.*, 2005 for further information on the mechanism of 4-AP). However, the mechanisms responsible for this response to stimulation have not yet been fully investigated.

E-mail: grainne.mcentee@virgin.net

The timing and expression of the MRFs are important for the proper pattern of myogenesis to occur during development. The MRFs are considered as the regulators of muscle phenotype, previous findings indicating the differential accumulation of MyoD and myogenin in fast and slow muscle respectively as fibre type transformation is induced. Characteristically, slow fibres possess relatively more MRF4 than fast fibres (Walters *et al.*, 2000). In addition, MyoD and Myf5 exhibited a greater fold increase in the fast muscle in response to denervation (Walters *et al.*, 2000). Because of the importance of innervation to muscle development, the nAChRs can be considered as playing an indirect role in the myogenic pathway. Acetylcholine receptors can be regarded as a measure of the degree of innervation to which muscle is subject. The nAChR and its subunits form part of the skeletal muscle surface, centrally involved in signal transmission at the neuromuscular junction (O'Reilly *et al.* 2003). However, despite the body of research on gene expression associated with neuromuscular development, the effect of pre-natal neuromuscular stimulation on gene expression of myogenic and neurogenic factors and the subsequent impact of a change in muscle gene expression on post-hatch skeletal muscle have not yet been fully explored.

In the following paper, we aim to understand the effect of increased pre-hatch neuromuscular stimulation in embryonic chick muscle on the expression of some key factors involved in muscle development (i.e. nAChR, myogenin, myostatin (GDF-8), MyoD and MRF4) in an attempt to elucidate their role in fast and slow muscle development. Our hypothesis is that stimulation of movement *in ovo* results in an altered expression of the genes associated with the neuromuscular system, which may alter post-hatch skeletal muscle development.

Materials and methods

ANIMALS

148 fertile White Leghorn eggs (Joice & Hill Poultry Ltd., Norfolk, UK) were incubated in a static air incubator (Brinsea Products Ltd., Somerset, UK) at 37.5°C ± 0.5° and relative humidity of approximately 65-70%. Eggs were turned 4 times daily (every 3-4 hours) to promote normal growth and development. On embryonic day 9, the eggs were windowed above the air sac and sealed with clear tape (5mm diameter). Following the windowing procedure, the eggs were divided into stimulated (S) and control (C) groups. The S group received 100μl of sterile PBS buffer containing 0.1 mg 4-AP on embryonic days 10 and 11, while the embryos in the C group received 100μl of PBS only. On embryonic days 12 and 13 of incubation, the S group were injected with 100μl of PBS buffer containing 0.2 mg 4-AP and the embryos in the C group were administered 100μl of PBS only. The window was re-sealed with tape and the eggs returned to the incubator. The concentration of 4-AP used was based on the drug's half-life and on work done by Osborne (2000) in which a stimulatory effect on movement was observed. Table 1 summarises the number of embryos used and the mortality over the course of the experiment, which was significantly reduced from our previous experiment (Heywood *et al.*, 2005).

SAMPLING

Four or five embryos on each sample day (E15 to E20) from each group (C and S) were

**Table 1. Mortality rates and total sample size record of the control and 4-AP stimu-
lated embryos.**

Group	Number of eggs	Infertile/did not develop before day 10 (both groups)	Mortality > day 10	Total % mortality in C and T embryos
Control	65	12	8	24%
Stimulated	83		16	

sacrificed by decapitation and the body mass of embryos (including the yolk sac) at all ages
was measured. On each day of sampling, the right and left anterior *latissimus dorsi* muscle
(ALD) and the right and left posterior *latissimus dorsi* muscle (PLD) were identified and
dissected from each embryo. The muscles from the right side were wrapped in ethanol-
sprayed foil and snap frozen immediately in liquid nitrogen for analysis of expression of
myogenic and neurogenic mRNA transcripts.

RNA EXTRACTION & CDNA SYNTHESIS

Total RNA was extracted using tri-reagent as recommended by the supplier (Sigma-Aldrich,
UK). The RNA was dissolved in RNAse-free water and was run on an ethidium bromide gel
to ensure that there was no degradation. The RNA samples were purified using the RNeasy
purification kit with DNAse treatment to remove genomic DNA contamination following
the manufacturer's recommended protocol (Qiagen, UK) and were quantified using gene
spectrophotometry (Naka Instruments, Japan) to measure the concentration of total RNA.

Using the Omniscript kit (Qiagen, Crawley, UK), reverse transcription was performed
on the samples to produce cDNA according to the manufacturers' instructions. The cDNA
was stored at -20°C for analysis by real-time PCR.

REAL TIME QUANTITATIVE PCR

RT-PCR was performed using LightCycler technology (Roche Diagnostics) as described by
Hameed *et al.*, (2002). The primers used for real time PCR were designed using the Primer3
web interface (Rozen and Skaletsky, 2000) and were synthesized by MWG (Ebersberg,
Germany). The sequences of the primers used are given in Table 2. The SYBR Green PCR
kit was used and the reactions amplified using the DNA Opticon Engine (MJ Research) as
recommended by the manufacturers. Serial dilutions of standard DNA of known concentrations were included in each run from which a standard curve was created and used to quantify concentration of DNA. Target specificity was further confirmed by running samples on
an agarose gel. PCR runs which yielded a standard curve with an *r*-value greater than 0.95
were used in the analysis.

STATISTICAL ANALYSIS

All data sets were checked using residuals to ensure a normal distribution of the data prior to
statistical analysis. Data that deviated from a normal distribution were log-transformed

Table 2. Forward and reverse primer sequences used in RT-PCR analyses.

Primer Name	Sequence (5' to 3')	Product Size (bp)	Accession No.
nAChR for	GTT CCT GGC TTT CCC CTA AC	191	J05642
nAChR rev	AGC GGG AGA AGA GTT TGT GA		
MRF4 for	CAG GAC AAA ATG CAG GAG GT	224	XM416115
MRF4 rev	GAG GAA ATG CTG TCC ACG AT		
MyoD for	ACT ACA CGG ATT CAC CAA ATG	146	AY641567
MyoD rev	CCC TTC AGC TAC AGC TTC AGC		
Myostatin for	AGG TGA AGA TGG ACT GAA CC	143	AF019625
Myostatin rev	CTT CAA AAT CCA CTG TCA GC		
Myogenin for	CAG AGG TTT TAC GAT GGG AA	291	D90157
Myogenin rev	CAG AGT GCT GCG TTT CAG AG		

prior to analysis of variance (ANOVA) and are presented as log copy number in the tables and figures below. Data were analysed for treatment and treatment*day interaction effects using the general linear model in Minitab Release Software 12.21 (1998). Where an interaction effect was found, the groups were analysed separately to pinpoint statistical differences between days within an experimental group. Statistical significance was verified using Tukey's pairwise comparisons. The level of statistical significance was taken as $P < 0.05$. Data are presented as means ± S.E.M.

Results

NACHR AND MYOGENIC GENE EXPRESSION IN SLOW MUSCLE

The results for the mRNA transcript expression of nAChR expression in the slow muscle are presented in table 3 and figure 1(a). There was significantly more nAChR expressed in the control compared to the stimulated group in the ALD muscle. There were no treatment effects apparent on any sample days with the exception of ED18 on which there was a significant suppression of nAChR expression in the ALD muscle of the stimulated compared to the control group. The mRNA expression levels of the myogenic regulatory factors in the ALD muscle are presented in table 3 and figures 1(b) – 1(e) respectively. Analysis for treatment and treatment x day interaction effects indicates that there was significantly more myogenin mRNA expressed in the control compared to the stimulated ALD group on ED15 only. There was significantly more total myostatin mRNA expressed in the ALD muscle of the stimulated compared to the control group. There were no significant differences in the expression of MRF4 or MyoD mRNA in the control compared to the stimulated groups of the slow muscle on any of the sample days.

NACHR AND MYOGENIC GENE EXPRESSION IN FAST MUSCLE

The results indicated a significant up-regulation in the mRNA expression of nAChR in the PLD muscle of the stimulated compared to the control group (Table 4). The up-regulation in nAChR mRNA expression occurred on ED17 and 18 (Figure 2(a)). Myogenin mRNA expression in the PLD muscle was significantly higher in the control compared to the stimulated group, specifically on ED16 and ED20 (Table 4 and figure 2(b) - 2(e)). There were significantly higher levels of myostatin mRNA expressed in the PLD muscle of the stimulated group on embryonic day 17.

Table 3: Real time RT-PCR analyses on total RNA extracted from the ALD muscle from pooled data ED15-ED20. Results are presented as log-transformed means ± SEM. P-values are also presented (*n=32*).

| Gene | ALD Log Copy Number | | P-value |
	Control	Stimulated	
nAChR	26.3 (0.1)	25.7 (0.2)	0.03
Myogenin	27.0 (0.1)	26.1 (0.1)	0.05
Myostatin	17.4 (0.2)	18.2 (0.2)	0.01
MRF4	27.6 (0.1)	27.5 (0.1)	NS
MyoD	17.5 (0.1)	17.5 (0.1)	NS

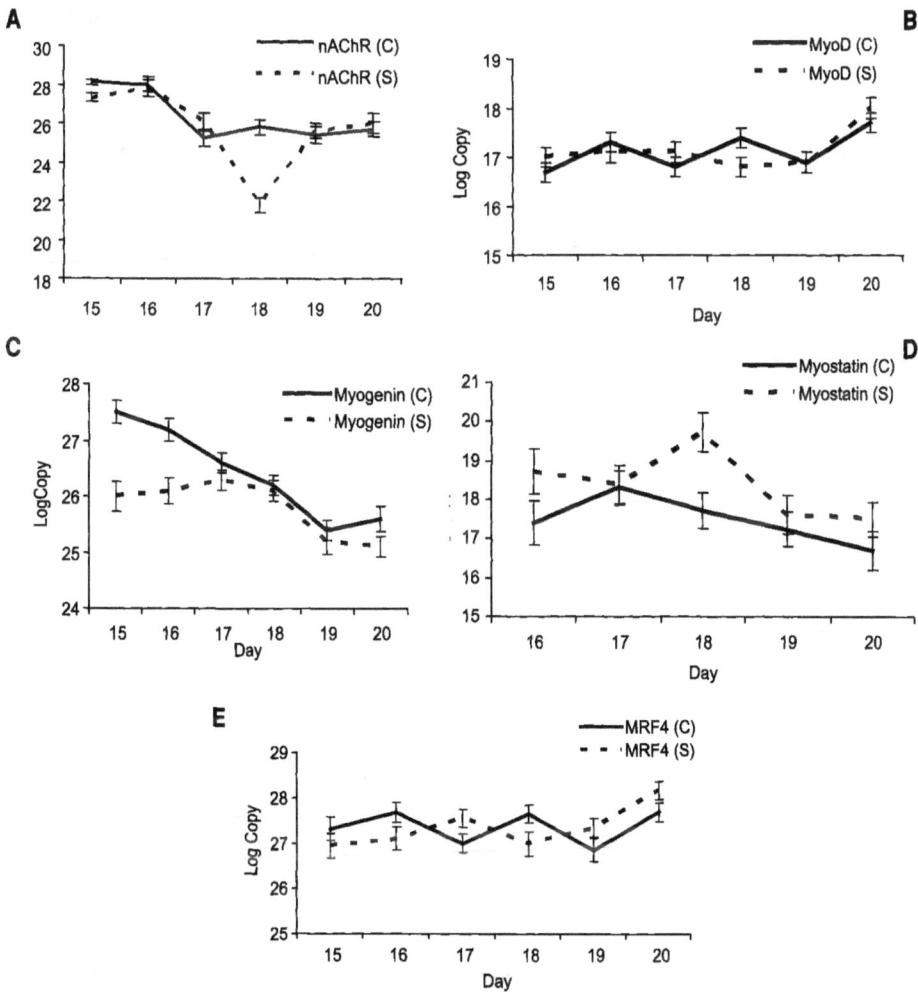

Figure 1(a) – 1(e): Real time RT-PCR analyses on total RNA extracted from the ALD muscle separated to show gene expression levels across time from ED15-ED20. Control group is solid line; Stimulated group is broken line. X-axis represents embryonic day; Y-axis represents log copy number. Results are presented as log-transformed means ± SEM (*n=4-6* per treatment group).

Table 4: Real time RT-PCR analyses on total RNA extracted from the PLD muscle from pooled data ED15-ED20. Results are presented as log-transformed means ± SEM. P-values are also presented ($n=32$).

Gene	Control	PLD Log Copy Number Stimulated	P-value
nAChR	24.9 (0.1)	25.8 (0.1)	0.001
Myogenin	34.1 (0.2)	33.1 (0.2)	0.02
Myostatin	22.7 (0.2)	24 (0.2)	0.001
MRF4	27.5 (0.1)	27.6 (0.1)	NS
MyoD	17.1 (0.06)	17.3 (0.06)	NS

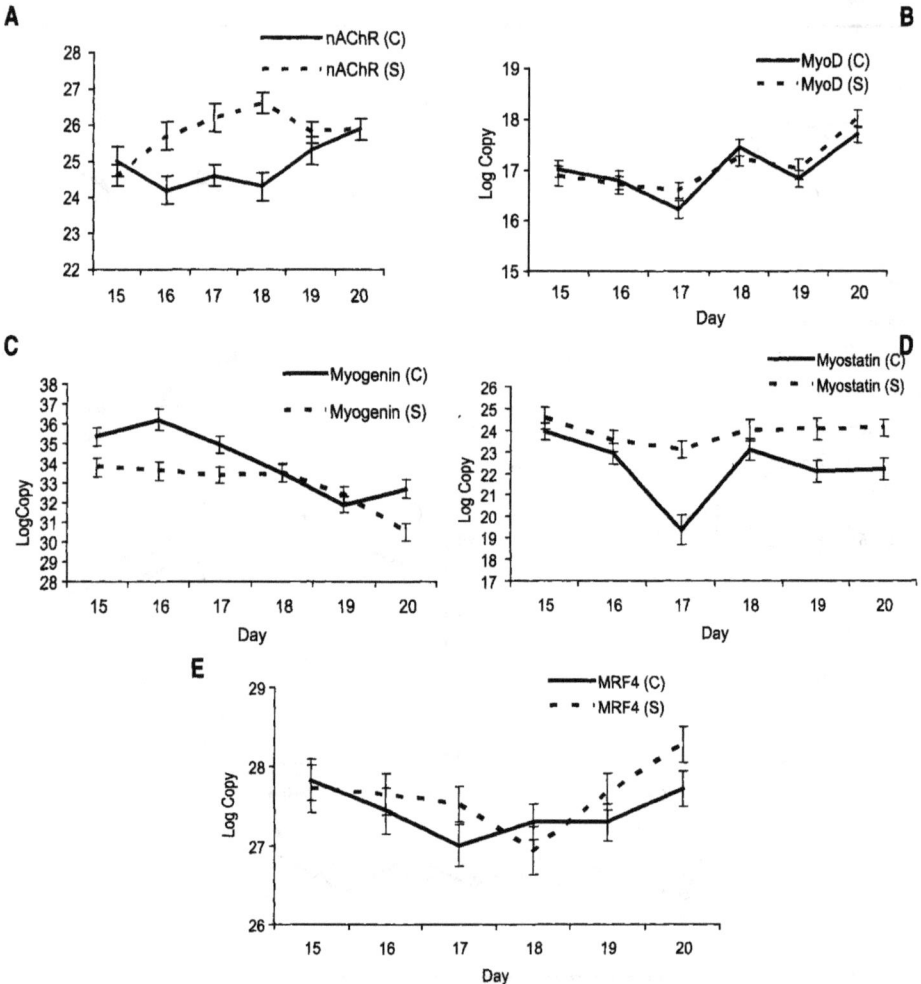

Figure 2(a) – 2(e): Real time RT-PCR analyses on total RNA extracted from the PLD muscle separated to show gene expression levels across time from ED15-ED20. X-axis represents embryonic day; Y-axis represents log copy number. Control group is solid line; Stimulated group is broken line. Results are presented as log-transformed means ± SEM ($n=4$-6 per treatment group).

Discussion

Development of the neuromuscular junction, which represents the interface between motorneurons and skeletal muscle, occurs between ED7 and 10 in the chick (Whittow, 1999). Just before the onset of motorneuron death at ED 6.5-7, the motorneuron axons reach their target muscles in the hindlimb (Dahm and Landmesser, 1988). By ED8, after the nerve has grown further into the muscle, pronounced visible movements of the hindlimb can be observed, and are produced by spontaneous motorneuron activity. In the normal developing embryo, this spontaneous neuronal activity is highest between ED10 to ED13. One method by which neuromuscular stimulation can be achieved is through the application of 4-AP, a K$^+$ channel blocker known to enhance transmitter release from motor nerve endings in both adult and neonatal rats (Lundh and Thesleff, 1977; Dekkers, *et al.*, 2001). The protective effect of 4-AP on nerve terminals has been previously observed in post-natal animals, in which the reduced K$^+$ transients present during the withdrawal period of synaptic input resulted in the preservation of many nerve terminals that would otherwise have lost contact with the muscle (Evers, 1987).

The development and long-term survival of skeletal muscle depends on innervation. Fibre type is considered as being determined by innervation (Redenbach *et al.*, 1988; Wigmore and Evans, 2002) and depending on the target fibre type, innervation can be multiple (in the case of Type I fibres) or focal (in the case of Type II fibres) (Ovalle *et al.*, 1999). As such, it is possible that stimulation of the muscle prior to commitment of the myoblasts to differentiation could effectively alter post-natal skeletal muscle development. While this has previously been demonstrated in embryonic muscle immediately pre-hatch (Heywood *et al.*, 2005), it has yet to be established if such manipulation during development impacts post-hatch muscle growth and development via changes in the expression of key neurogenic and myogenic factors.

We studied the expression of four key differentiation genes namely, myogenin, MyoD, MRF4 and GDF-8 (i.e. myostatin), a negative regulator of myogenic proliferation as well as that of nAChR, directly associated with muscle innervation and which has a profound effect on skeletal muscle development. We had previously proposed that increased fibre number was as a result of prolonged polyneuronal innervation (Heywood *et al.*, 2005). The results of the present study indicated an up-regulation of acetylcholine receptor expression on ED17 and 18 in the stimulated PLD group compared to the control, indicating that nAChR mRNA levels are highest during the latter stages of secondary myogenesis. Phillips *et al.*, (1985) established the presence of nAChR clusters on chick ALD muscle by embryonic day 10-14. The application of neuromuscular stimulation may have delayed motorneuron withdrawal or promoted polyneuronal innervation, suggested by the greater nAChR mRNA expression in the stimulated group during late development, when axonal withdrawal is occurring. However, a different response is observed in the slow muscle phenotype, suggesting that muscle fibre type has strict control over the mechanism of innervation and can protect itself against prolonged stimulation by controlling the nAChR expression. The altered expression of these factors therefore may arise from the altered expression of the MRFs. It is possible that there is a signalling mechanism by which the myogenic program "tells" the nervous system of its requirements. A reduced expression of certain myogenic factors may trigger the increased expression of neurogenic factors. Greensmith *et al.*, (1996) have speculated that prenatal induction of transmitter release may enhance the maturation of fast fibres, enabling in-growing axons to be more successful in establishing neuromuscular contacts.

In the early stages of embryonic development, MyoD activates the muscle-specific genes that determine muscle phenotype. MyoD and its relatives are important transcriptional factors involved in the removal of the myoblast from the cell cycle. The results of the present study suggest that there is a significant down-regulation of myogenin mRNA expression on day 16 and day 20 pre-hatch in the PLD muscle of the stimulated compared to the control group. In comparison, myogenin expression is significantly depressed on embryonic day 15 only in the ALD of the stimulated compared to the control group. The depression of myogenin mRNA expression in the PLD as a result of neuromuscular stimulation may indicate that myoblast proliferation was prolonged and possibly differentiation was partially arrested in the stimulated PLD muscle.

MRF4 is involved in the control of terminal differentiation of myofibres however, the stimulation does not appear to affect the expression of MRF4 mRNA positively or negatively in either muscle fibre phenotype. This is supported by a study by Walters *et al.*, (2000) in which there were no observable differences in the expression of MRF4 in the PLD muscle due to neuromuscular stimulation. This can in part be explained by their findings that MRF4 is important in the maintenance of an established slow muscle phenotype. A previous study has found that MyoD expression was concomitant with the onset of nAChR transcription (Charbonnier *et al.*, 2003), a result that confirms the previous findings of Piette *et al.*, (1990) when they located two MyoD-binding sites on the DNA adjacent to an alpha subunit gene of the chick muscle AChR. In the present experiment, with respect to the slow muscle phenotype, there was a significant down-regulation of nAChR on ED 18. This result would support the finding that innervation inhibits nAChR expression, but this only appears to be the case in the slow muscle phenotype. Conversely, in the PLD muscle, an up-regulation of nAChR expression is observed in the stimulated muscle with no significant increase observed in MyoD expression. In light of these results, it would appear that the activation of the nAChR alpha-subunit implies a non-specific effect of the different MRFs that depends on the predominant muscle fibre type.

Myostatin is a negative regulator of myoblast differentiation and in line with the fact that the stimulated muscle is undergoing a prolonged proliferative phase of myoblast formation, a significant increase in mRNA expression of myostatin is observed on ED18 in the ALD muscle. However, in the PLD muscle, its up-regulation is observed and follows similar treatment-induced differences in nAChR mRNA expression. The results would suggest that the stimulatory treatment, which has enhanced nAChR expression in the PLD muscle, has led to an increased proliferation resulting in more myoblast differentiation. Effectively, more myoblast differentiation would require a higher level of myostatin expression. It is possible that the up-regulation of myostatin observed in the stimulated ALD muscle on ED18, may have contributed to the simultaneous depressive effect on nAChR expression. The neuromuscular stimulation that caused this increase in the level of myostatin expression may effectively offer a feedback signal to the nAChR to reduce their activity, thereby protecting the slow muscle from the over-proliferative effects induced by neuromuscular stimulation.

In conclusion, the differential regulation of myogenic and neurogenic molecular factors in fast and slow muscle indicates that it is the change in their expression during *in ovo* stimulation that likely contributes to an increase in muscle fibre number in the post-hatch phenotype. There are clear relationships between the regulation of transcripts that have been stimulated *in ovo* present in the motorneuron, the skeletal muscle and the interface between the two (NMJ), which are dependant on the muscle type. Pre-natal stimulation can alter the expression of these transcripts which in turn, may affect the phenotype of the muscle post-hatch. In the current experiment, we have attempted to show the fibre-type specific responses of muscle to neuromuscular stimulation as well as the changes in the timing of gene expression that are elicited by neuromuscular stimulation in the embryo.

References

BAYOL, S., JONES, D., GOLDSPINK, G. and STICKLAND, N.C. (2004). The influence of undernutrition during gestation on skeletal muscle cellularity and on the expression of genes that control muscle growth. *British Journal of Nutrition* **91** (3): 331:339.

CHARBONNIER, F., DELLA GASPERA, B., ARMAND, A-S., LE´COLLE, S., LAUNAY, T., GALLIEN, C-L and CHANOINE, C. (2003) Specific activation of the acetylcholine receptor subunit genes by myoD family proteins. *Journal of Biological Chemistry* **278**(35): 33169–33174.

DAHM, L.M. and LANDMESSER, L.T. (1988) The regulation of intramuscular nerve branching during normal development and following activity blockade. *Developmental Biology* **130**(2):621-44.

DEKKERS, J., WATERS, J., VRBOVA, G. and GREENSMITH, L. (2001). Treatment of the neuromuscular junction with 4-aminopyridine results in improved reinnervation following nerve injury in neonatal rats. *Neuroscience* **103**(1): 267-74.

EFTIMIE, R., BRENNER, H.R. and BUONANNO, A. (1991) Myogenin and MyoD join a family of skeletal muscle genes regulated by electrical activity. *Proceedings of the National Academy of Science* USA. **88**(4):1349-53.

EVERS, J.V. (1987) *The reorganisation of synaptic inputs to developing mammalian skeletal muscle*. Ph. D. Thesis, London University.

GREENSMITH, L., DICK, J., EMANUEL, A.O. and VRBOVA, G. (1996) Induction of transmitter release at the neuromuscular junction prevents motorneuron death after axotomy in neonatal rats. *Neuroscience* **71**(1): 213-220.

HAMEED, M., ORRELL, R.W., COBBOLD, M., GOLDSPINK, G. and HARRIDGE, S.D. (2003) Expression of IGF-I splice variants In young and old human skeletal muscle after high resistance exercise. *Journal of Physiology* **547** (Pt 1):247-54.

HEYWOOD, J.L., G. M. MCENTEE and N. C. STICKLAND (2005) *In ovo* neuromuscular stimulation alters skeletal muscle phenotype in the chick. *Journal of Muscle Research and Cell Motility*. (In press).

LUNDH, H. and THESLEFF, S. (1977) The mode of action of 4-aminopyridine and guanidine on transmitter release from motor nerve terminals. *European Journal of Pharmacology* **42**: 411-412.

MALTBY, V., SOMAIYA, A. FRENCH, N.A. and STICKLAND, N.C. (2004) *In ovo* temperature manipulation influences post-hatch muscle growth in the turkey. *British Poultry Science* **45** (4): 491-498.

MINITAB (1998). *Release 12.1 for Windows*. Minitab Statistical Software.

OVALLE, W.K., DOW, P.R. and NAHIRNEY P.C. (1999) Structure, distribution and innervation of muscle spindles in avian fast and slow skeletal muscle. *Journal of Anatomy* **194** (3):381-94.

O'REILLY, C., PETTE, D. and OHLENDIECK, K. (2003) Increased expression of the nicotinic acetylcholine receptor in stimulated muscle. *Biochemical Biophysical Research*

Communication. **300** (2):585-91.

OSBORNE, A. C. (2000). *Mechanisms by which movement exerts its essential role on diarthordial joint cavity.* PhD Thesis.

PIETTE, J, BESSEREAU, JL, HUCHET, M. and CHANGEUX, JP. (1990) Two adjacent MyoD1-binding sites regulate expression of the acetylcholine receptor alpha-subunit gene. *Nature* **345** (6273):353-5.

PHILLIPS, W.D., LAI, K. and BENNETT, M.R. (1985) Spatial distribution and size of acetylcholine receptor clusters determined by motor nerves in developing chick muscles. *Journal of Neurocytology* **14**(2):309-25.

REDENBACH, D.M., OVALLE, W.K. and BRESSLER, B.H. (1988) Effect of neonatal denervation on the distribution of fiber types in a mouse fast-twitch skeletal muscle. *Histochemistry* **89**(4):333-42.

ROZEN, S. and SKALETSKY, H. (2000) Primer3 on the WWW for general users and for biologist programmers. *Methods in Molecular Biology* **132**:365-86.

TRACHTENBERG, J.T. (1998) Fiber apoptosis in developing rat muscles is regulated by activity, neuregulin. *Developmental Biology* **196**(2):193-203.

WALTERS, E.H., STICKLAND, N.C., and LOUGHNA, P.T. (2000) The expression of the myogenic regulatory factors in denervated and normal muscles of different phenotypes. *Journal of Muscle Research and Cell Motility* **21**(7): 647-53.

WHITTOW, J.C. (1999) *Sturkie's Avian Physiology.* Academic Press.

WIGMORE, P.M. and EVANS, D.J. (2002) Molecular and cellular mechanisms involved in the generation of fibre diversity during myogenesis. *Int. Rev. Cytol.* **216**:175-232.

Characterization of neuronal hypothalamic plasticity in the chicken: a comparative analysis

NAGARAJA SALLAGUNDALA,[1] KRASSIMIRA YAKIMOVA[2] AND BARBARA TZSCHENTKE[1]

[1]Humboldt-Universität zu Berlin, Institut für Biologie, AG Perinatale Anpassung, Philippstr.13, D-10115 Berlin, Germany; [2]Department of Pharmacology, Faculty of Medicine, Medical University, 1431 Sofia, Bulgaria

Characterization of neuronal hypothalamic plasticity in chicken brain slices aims to investigate the influence of age on thermosensitivity in the preoptic area of the anterior hypothalamus (PO/AH) by extracellular recordings in the age group of 5, 10, 15, 20 and 30 days old birds. Firing rate of neuronal activity was recorded extracellularly during sinusoidal temperature changes. Investigations reveal high proportion of cold-sensitive neurons compared to the warm and insensitive neurons in all the age groups studied, which has not been found in literature so far. In chickens cold sensitivity shows an increase from day 5 until day 20 and is as high as 52 % in 20 days old birds. Between day 20 and day 30 neuronal cold sensitivity again shows a major shift and falls to 34%. In mammals as well as in adult Pekin ducks the cold – sensitive neurons were less than 10 %. Between birds a species specificity of the early development of neuronal hypothalamic thermosensitivity could be clearly demonstrated. Hence the high hypothalamic cold sensitivity seems to be a specific characteristic in juvenile birds. It suggests a possible thermoregulatory role of cold – sensitive neurons in the chicken species in the observed age groups. Existence of inherent nature of cold – sensitive neurons is a question for investigation, hence synaptic blockade of cold-sensitive neurons were performed, which shows inherent tendency to a certain degree. Further, study with GABA receptor agonist baclofen assumes significant changes in neuronal characteristics. Baclofen exhibited inhibitory action prominently on cold – sensitive neurons. GABA (B) mechanisms also increased thermal coefficient of the cold-sensitive PO/AH neurons.

Key words: Hypothalamic neurons, temperature sensitivity, chicken, brain slices, GABA, cold sensitivity, synaptic blockade

Introduction

The goal of temperature regulation in homoeothermic animals is the maintenance of a stable body core temperature under most conditions. To realize this, the thermoregulatory system employs all of the systems of the body and integrates their activities into appropriate and

*Corresponding author: Barbara Tzschentke
E-mail: Barbara.Tzschentke@rz.hu-berlin.de

coordinated reactions. For instance, the effect of feed on maintaining the balance between heat production and heat loss is probably the most direct effect of high ambient temperature in poultry. This means that a decrease in growth or egg production in a hot environment is mainly caused by a reduction of feed intake in order to reduce heat production (MacLeod, 1981; Waibel and MacLeod, 1993; Shane, 2001). Under the long-term influence of specific environmental temperatures the organisms develop various adaptation mechanisms to minimize the effect of the respective temperature (acclimatisation). A very sensitive phase is the prenatal and early postnatal development. During critical phases environmental influences can induce long-lasting changes in the development of physiological control systems, like the thermoregulatory system (epigenetic temperature adaptation, Nichelmann *et al.*, 1994; Tzschentke and Nichelmann, 1997; Tzschentke *et al.*, 2004; thermal conditioning, Yahav and McMurtry, 2001). During acclimatisation as well as epigenetic temperature adaptation changes occur in peripheral mechanisms (e.g. heat production, heat loss mechanisms) as well as in the central nervous mechanisms (e.g. neuronal sensitivity of the central controler) (Tzschentke and Nichelmann, 1997).

In birds as well as in mammals, the hypothalamus plays a central role in thermoregulation and acts as a coordinating centre influencing the different effector areas. The hypothalamic area has thermosensitive neurons. These neurons react on the peripheral as well as on local temperature changes with remarkable differences in firing rate. These neurons accordingly are classified as warm-sensitive (WS), cold-sensitive (CS), and temperature-insensitive (TI) neurons. The neuronal hypothalamic thermosensitivity shows a high plasticity in adults. This kind of plasticity has been exhibited in prenatal and also in juvenile mammals and birds. Neuronal hypothalamic thermosensitivity can be influenced by different factors, like hormones (Moravec and Pierau, 1994), neuropeptides (Schmid *et al.*, 1993), neurotransmitters (Yakimova *et al.*, 1996, 2005), Ca^{++} (Schmid and Pierau, 1993), environmental factors, like temperature adaptation (Pierau *et al.*, 1994; Tzschentke *et al.*, 1994). Basic knowledge of the development of hypothalamic neuronal plasticity in poultry species is lacking. But it is a prerequisite to understand central cellular mechanisms, e.g. of temperature adaptation. Investigations on neuronal hypothalamic plasticity in brain slices of birds have been well established in earlier investigations of our laboratory in Muscovy ducks. Age related changes of the neuronal hypothalamic thermosensitivity, especially during the early development, could be clearly shown in embryos as well as growing Muscovy ducklings (Tzschentke and Basta, 2000). It seems that in birds during early development hypothalamic CS neurons play a prominent role in the central network, which controls body temperature. Further, in this species the neuronal sensitivity of the hypothalamus was changed by neuropeptides (Tzschentke *et al.*, 2000) as well as epigenetic temperature adaptation (Tzschentke and Basta, 2002; Tzschentke *et al.*, 2004).

The present investigation aims to characterize the neuronal hypothalamic plasticity with special focus on neuronal cold sensitivity during early post hatching development in another poultry species, the chicken (*Gallus gallus f. domestica*), using the method of extracellular recordings in brain slices. Age related development in neuronal cold sensitivity and the existence of inherent cold sensitive neurons have been addressed in the present study. In the light of these investigations, the study of GABA – the main inhibitory neurotransmitter in the brain, which is also prominently involved in the neuronal thermosensitivity, assumes significance for the present probe. Comparative aspects of neuronal hypothalamic thermosensitivity between mammals and different bird species will be discussed, in the light of the present studies.

Materials and methods

Experiments were carried out in 10 to 30 days old White Leghorn chicks (*Gallus gallus f. domestica*). On the day of the experiments the chicks were decapitated and the brain was removed. Slices (400 μm) including the preoptic area of the anterior hypothalamus (PO/AH) were prepared and incubated for 2 hours in oxygenated (95% O_2, 5% CO_2) artificial cerebrospinal fluid (ACSF) at 35°C. The experimental setup was described by Schmid *et al.* (1993). For the extracellular recordings the slices were transferred into the recording chamber, continuously perfused with oxygenated ACSF at a rate of 2.5 ml/min. The basic temperature in the recording chamber was kept constant at 40°C (this bath temperature approximately corresponds to the deep body temperature of juvenile chicks, Tzschentke and Nichelmann, 1999). For the extra cellular recordings of the neuronal activity glass electrodes were used. Recordings were carried out from spontaneously active neurons in the area located between *commissura anterior* and *chiasma opticum* lateral to the third ventricle (PO/AH). Neuronal activity and slice temperature were recorded by conventional electrophysiological equipment and stored on a personal computer using a 1401 interface (Cambridge Electronic Design (CED) and the CED software Spike 2. Only activity of single units, which were separated using a window discriminator, were stored on the computer. For identification of thermosensitivity of the single neurons, temperature was sinusoidally altered between 37-43°C by means of a Peltier thermoassembly (rate of temperature changes 0.02°C/s). Temperature sensitivity was calculated by a computer program relating the discharge rate of the neuron (bin width = 5 s) to the actual temperature, and fitting either one linear or two piecewise regression lines through the data. The slope of the steepest regression line was used as the temperature coefficient (TC) of the unit (Vieth, 1989). Temperature-sensitive neurons were defined by a TC ≥0.6 imp/s/°C with a positive sign for WS neurons and a negative sign for CS neurons; all other neurons were TI by this definition (Nakashima *et al.*, 1987). Changes in neuronal tonic activity (firing rate) were calculated with the aid of the same computer program, providing information on the mean value of firing rate for the duration of 1 min, recorded just prior to each temperature stimulus.

On some CS neurons the modulatory action of GABA(B) agonist was tested. The GABA(B)-agonist R(+)-baclofen hydrochloride (1 mcM) (Sigma-Aldrich Chemie GmbH) was diluted in ACSF just prior to the application. Before application of the test substance, the temperature sensitivity of a given neuron was determined using two sinusoidal temperature stimuli at intervals of 5 minutes. Superfusion of test substance was started not before 5 min after the last control temperature stimulus; baclofen was applied for 5 min before the next temperature stimulus was performed. Superfusion returned to ACSF after this stimulus was completed and a further temperature stimulus was given after a delay of at least 10 min. Additional temperature stimuli were applied in anticipation of complete recovery. Only one neuron per slice was tested.

To show that neuronal hypothalamic cold sensitivity can also be an inherently neuronal function, in some CS neurons the influence of synaptic blockade with Ca^{++}-free ACSF was tested. Sinusoidal temperature changes were applied before, during and after superfusion with Ca^{++}-free ACSF. The experimental protocol was similar to that used for GABA application.

For characterization of the neuronal hypothalamic thermosensitivity the proportion of WS, CS and TI neurons in the PO/AH was determined in relation to all neurons investigated (Tzschentke and Basta, 2000). For instance, an increase in the proportion of CS neurons and

a decrease in WS neurons in relation to all neurons investigated was used as a sign of elevation in total neuronal cold-sensitivity of the PO/AH.

STATISTICAL EVALUATION

For statistical evaluation, chi-square (χ^2)-test was employed to test for differences between
age groups. The data on influence of GABA on CS neurons has been presented as means ±
S.E.M. For statistical evaluation a paired t-test was performed. A p - value of <0.05 was
considered statistically significant.

Results and discussion

AGE RELATED CHANGES IN NEURONAL HYPOTHALAMIC THERMOSENSITIVITY

Characterization of temperature sensitive and insensitive neurons in 5, 10, 15, 20 and 30-
days-old birds has been performed. Table 1 summarizes the number of investigated neurons
in each class of sensitivity as well as the proportion (percentage) of WS-, CS- and TI- neurons in the PO/AH, determined in relation to all neurons investigated in each age group.
Within the classes of thermo sensitive neurons the predominance of cold sensitive neurons
has been observed in all investigated age groups.

**Table 1. Type, number and percentage (in relation to all neurons investigated in the
respective age group) of neurons of the anterior hypothalamus investigated in 5- to 30-
days old chickens.**

Age groups	Type of neurons	Number of neurons	Percentage of neurons investigated in relation to all neurons in each age group
5 days	Warm sensitive	15	22
	Cold sensitive	24	35
	Insensitive	29	43
10 days	Warm sensitive	14	19
	Cold sensitive	23	32
	Insensitive	35	49
15 days	Warm sensitive	17	21
	Cold sensitive	34	43
	Insensitive	29	36
20 days	Warm sensitive	13	18
	Cold sensitive	38	52
	Insensitive	22	30
30 days	Warm sensitive	24	29
	Cold sensitive	28	34
	Insensitive	30	37

For statistical evaluation, chi-square (χ^2)-test was employed to test for the differences between different age groups in warm, cold and insensitive neurons. A comparison was made
in case of WS, CS and TI-neurons between 5 days and 10 days; 10 days and 15 days; 15 days

and 20 days and finally between 20 days and 30 days. The significance levels are described accordingly.

In WS-neurons no significant levels have been exhibited in all the investigated age groups.

A comparison of CS-neurons between 5 days and 10 days; 10 days and 15 days; and 15 days and 20 days has been found to be non significant but highest order of significance was found when compared between 20 days and 30 days *viz.*, p<0.001.

A comparison of TI-neurons between 5 days and 10 days has been found to be non significant but a second degree of significance levels were found between 10 days and 15 days *viz.*, p<0.01; comparison between 15 days and 20 days, as well as between 20 days and 30 days, has been found to be non significant.

The neuronal hypothalamic thermosensitivity during early development in chicken is characterised by a high neuronal cold sensitivity. These findings are similar to that proved in Muscovy ducklings (Tzschentke and Basta, 2000). Figure 1 shows a comparison of this development in the chicken and Muscovy duck.

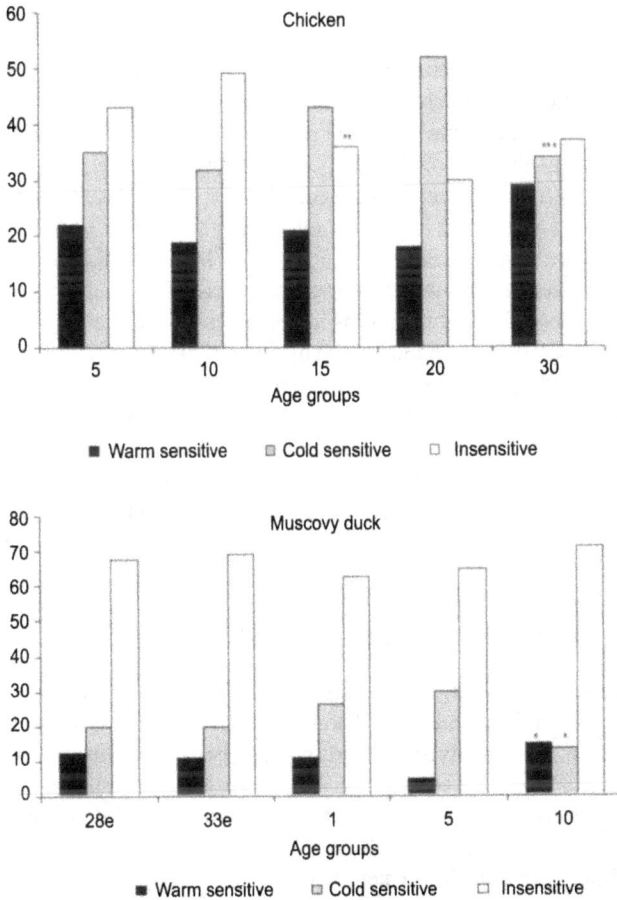

Figure 1. Influence of age on proportion of warm-, cold- and temperature-insensitive neurons in relation to all neurons investigated in their respective age groups in the preoptic area of the anterior hypothalamus of chicken and Muscovy ducks (data from Muscovy ducks published by Tzschentke and Basta, 2000). Asterisks represent significance at the level of *p<0.05, **p<0.01, ***p<0.001.

The neuronal hypothalamic thermosensitivity in Muscovy ducklings develops in two stages: *one* between days 28 of incubation and day 5 of post hatching, neuronal hypothalamic cold sensitivity increased from 20 to 30% whereas the proportion of WS neurons is low and decreased moderately from 12.5 to 5 % in relation to all neurons investigated; *two* between days 5 and 10 of post-hatching, neuronal cold sensitivity of the PO/AH decreased significantly to 14% whereas neuronal warm sensitivity increased significantly to 15%. However, in the Muscovy duck between days 5 and 10 after hatch, a qualitative change occurs in the development of neuronal hypothalamic thermosensitivity from the "juvenile" to the "adult" type. In adult Pekin ducks thermosensitivity of the PO/AH in brain slices is characterised by a low cold sensitivity (6.2%) and a high warm sensitivity (58.3%) (Nakashima *et al.*, 1987), which is similar to adult mammals (Boulant *et al.*, 1989). Similar high neuronal hypothalamic cold sensitivity during early ontogeny was also observed in prenatal temperature experienced Muscovy ducklings (Tzschentke and Basta, 2002). But this qualitative change from the "juvenile" to the "adult" type occurred in prenatal temperature experienced birds at an earlier stage of development (between day 1 and 5 of post hatching). Furthermore, this earlier change exclusively arose in the cold-sensitive neurons (Tzschentke *et al.*, 2004). In the actual study in chickens the first stage of the development of neuronal hypothalamic thermosensitivity, which is characterised by high and as well increasing neuronal cold sensitivity, occurs during a later developmental period. The increase in the percentage of CS neurons continues until the age of day 20 post hatching. There is also a shift from cold-sensitivity towards warm sensitivity in the second stage of development in neuronal hypothalamic thermosensitivity, which has been indicated by the results of the 30 days old birds. When we compare the results from chicken with the Muscovy ducks, we find a shift of hypothalamic thermosensitivity, which represents species specificity and is obviously related to the different developmental patterns in both species. Altogether, *Galliformes* during early postnatal development are characterised by a lower precocial status than *Anseriformes* (McNabb and Olson, 1996). When a comparative study is made between hatchlings of chicken and Muscovy ducks, a higher heat production capacity in Muscovy ducklings was observed, which enables them to stabilise the body temperature in a narrow temperature range (Tzschentke and Nichelmann, 1999). Because all effector activity is controlled by the nervous system, the earlier maturity of the duck brain corresponds with the developmental status of peripheral mechanisms.

MODULATORY ACTION OF GABA(B) AGONIST ON CS NEURONS

The thermoregulatory role of high neuronal hypothalamic cold sensitivity in birds during early development is an open question. The neuronal cold sensitivity seems to be a basic characteristic feature. As already described, it was also found after prenatal temperature manipulation.

In our investigations the influence of GABA(B)-agonist baclofen was tested in 11 CS-neurons of 10- to 15-days-old chickens. Under superfusion of baclofen, 9 of the 11 neurons (82%) were inhibited and 2 of the 11 neurons show an excitation. In comparison with WS and TI neurons, CS neurons show very clear reactions under pharmacological influence. A high number of CS PO/AH neurons respond to the GABA(B) receptor agonist baclofen. This indicates that these neurons express functional GABA(B) receptors. Also in the rat PO/AH a high percentage (97%) of neurons responded to baclofen (Yakimova *et al.*, 1996). For the whole population of investigated CS neurons a significant decrease in firing rate ($p<0.001$) and a significant increase in TC ($p<0.05$) was found in the present study. In WS and TI neurons the influence on firing rate was lower and no significant changes in the TC were observed (Yakimova *et al.*, 2005). In figure 2 the effect of GABA(B)-receptor agonist baclofen on a CS-neuron is shown. Recovery in firing rate and temperature sensitivity was observed after washout of the substance. The present results are preliminary studies in the direction of investigating cold sensitivity.

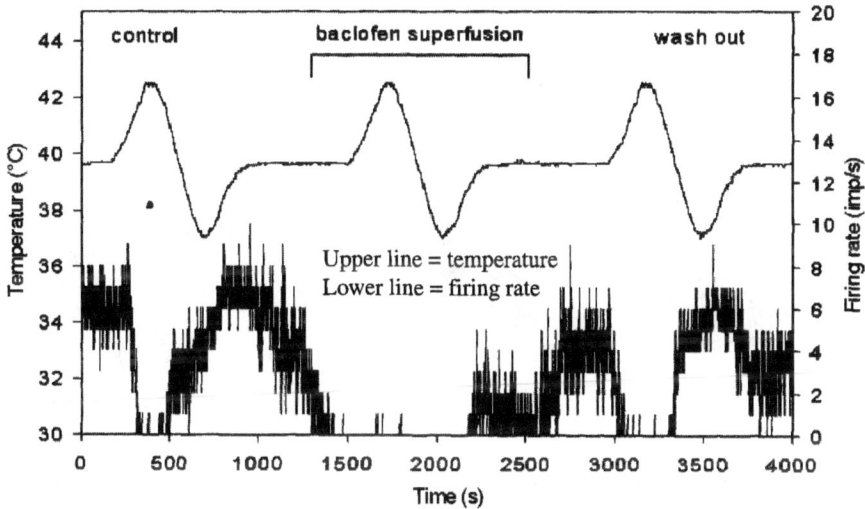

Figure 2. Effect of the GABA(B) receptor agonist baclofen on a cold-sensitive neuron in the preoptic area of the anterior hypothalamus of chicken: experimental protocol of the neuronal activity and temperature recorded close to the slice. Superfusion of baclofen (1 μM) completely stopped firing of the neuron. Note the recovery observed after washout of baclofen.

INHERENT CS NEURONS UNDER SYNAPTIC BLOCKADE

Existence of inherent nature of cold – sensitive neurons, which can act as thermosensors in the brain, is a question for investigation. Hence synaptic blockade of cold-sensitive neurons was performed. Whereas in adult mammals synaptic blockade shows that most synaptically driven WS neurons of the PO/AH are inherently thermosensitive, the inherent thermosensitivity of CS of PO/AH neurons discussed is controversial (Boulant and Dean, 1986). One hypothesis is that CS neurons in the hypothalamus can be viewed as interneurons inhibited by nearby warm sensitive neurons (Boulant *et al.*, 1989). On the other side, within the low number of CS neurons, which can be found in the mammalian as well as the hypothalamus of adult birds, single inherent CS neurons were found in brain slices during blockade with Ca^{++}-free/high-Mg^{++} -medium using extracellular recordings (Nakashima *et al.*, 1987). In another study using Ca^{++}-imaging and extracellular cell-attached patch recording, primary CS neurons with low threshold temperature were found in PO/AH of rats (Abe *et al.*, 2003).

    To test if inherently CS-neurons in the PO/AH exist, 12 CS neurons of 10- to 20-days-old chickens were investigated under synaptic blockade with Ca^{++}-free ACSF in the present study. Within the investigated CS neurons 3 increase the firing rate, and one of them strongly (Fig. 3), under cold load during synaptic blockade and show an inherently cold sensitivity. Superfusion with Ca^{++}-free ACSF induced an inhibition of the neuron. During sinusoidal temperature stimulation under synaptic blockade the neuron increased activity at the lowest temperatures around 37°C (TC: -1.83 imp/s/°C). After change to the control ACSF the neuron starts to recover.

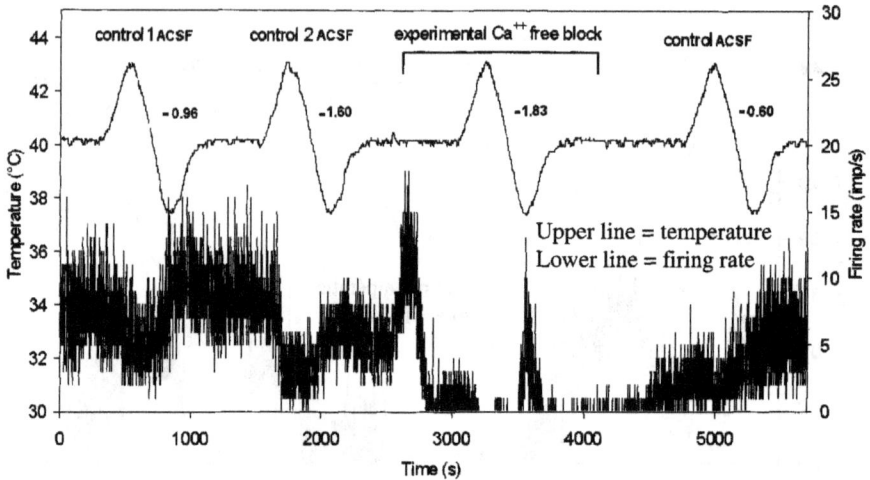

Figure 3. Example of cold sensitive hypothalamic neuron under synaptic blockade with Ca^{++}-free ACSF. The calculated thermal coefficient (TC) for a given temperature stimulus is indicated at the responses. Superfusion with Ca^{++}-free ACSF induced an inhibition of the neuron. During sinusoidal temperature stimulation under synaptic blockade the neuron increased activity at the lowest temperatures around 37°C (TC: -1.83 imp/s/°C). After change to the control ACSF the neuron starts to recover.

However, it is doubtful if Ca^{++} free ACSF removes all influences of presynaptic sensory neurons, where Ca^{++}- independent but voltage-dependent secretion occurs (Parnas *et al.*, 2000; Zhang and Zhou, 2002, Yang *et al.*, 2005). Additional investigations with specific receptor blockade are necessary for further identifications of primary CS neurons in juvenile birds.

In conclusion, the high neuronal hypothalamic cold sensitivity seems to be a specific characteristic feature of the early development in birds and possibly in mammals. A species specificity of the early development of neuronal hypothalamic thermosensitivity in birds could be clearly demonstrated. Primary (inherent) CS – neurons probably exist in the PO/AH of birds, which could act as thermo sensors.

References

ABE, J., OKAZAWA, M., ADACHI, R., MATSUMURA, K., and KOBAYASHI, S. (2003) Primary cold-sensitive neurons in acutely dissociated cells of rat hypothalamus. *Neuroscience Letters* **342:** 29-32.

BOULANT, J.A. and DEAN, J.B. (1986) Temperature receptors in the central nervous system. *Annual Reviews of Physiology* **48:** 639-654.

BOULANT, J.A., CURRAS, M.C. and DEAN, J.B. (1989) Neurophysiological aspects of thermoregulation: In: WANG. L.C.H. (Ed.) *Advances in Comparative and Environmental Physiology:* pp. 118-160. Springer-Verlag Berlin Heidelberg.

MACLEOD, M.G. (1981) Energy metabolism and the turkey. *Turkeys*, 26-33.

MCNABB, F.M.A. and OLSON J.M. (1996) Development of thermoregulation and its hormonal control in precocial and altricial birds. *Poultry & Avian Biology Reviews* **7**: 111-125.

MORAVEC, J. and PIERAU, FR.-K. (1994) Arginine vasopressin modifies the firing rate and thermosensitivity of neurons in slices of the rat PO/AH area. In: **PLESCHKA, K. and GERSTBERGER, R.** (Eds) *Integrative and cellular aspects of autonomic functions: temperature and osmoregulation*: pp. 143-152, John Linney Eurotext, Paris.

NAKASHIMA, T., PIERAU, FR.-K., SIMON, E. and HORI, T. (1987) Comparison between hypothalamic thermoresponsive neurons from duck and rat slices. *Pflügers Archiv* **409**: 236-243.

NICHELMANN, M., LANGE, B., PIROW, R., LANGBEIN, J. and HERRMANN, S. (1994) Avian thermoregulation during the perinatal period. In: **ZEISBERGER, E., SCHÖNBAUM, E. and LOMAX, P.** (Eds.) *Thermal Balance in Health and Disease. Advances in Pharmacological Science:* pp. 167-173, Birkhäuser Verlag AG, Basel.

PARNAS, H., SEGEL, L., DUDEL, J. and PARNAS, I. (2000) Autoreceptors, membrane potential and the regulation of transmitter release. *Trends in Neuroscience* **23**: 60-68.

PIERAU, FR.-K., SCHENDA, J., KONRAD, M. and SANN, H. (1994) Possible implications of the plasticity of temperature sensitive neurons in the hypothalamus. In: **ZEISBERGER, E., SCHÖNBAUM, E. and LOMAX, P.** (Eds) *Thermal balance in health and disease. Recent basic research and clinical progress:* pp. 31-36. Birkhäuser Verlag Basel.

SHANE, S.M. (2001) Enhancing production at high ambient temperatures. *World Poultry* **17**: 22-23.

SCHMID, H.A. and PIERAU, FR.-K. (1993) Temperature sensitivity of neurons in slices of the rat PO/AH area: effect of calcium. *American Journal of Physiology* **264** (*Regulatory Integrative Comparative Physiology* **33**): R440-R448.

SCHMID, H.A., JANSKÝ, L. and PIERAU, FR.-K. (1993) Temperature sensitivity of neurons in slices of the rat PO/AH area: effect of bombesin and substance P. *American Journal of Physiology* **264** (*Regulatory Integrative Comparative Physiology* **33**): R449-R455.

TZSCHENTKE, B. and BASTA, D. (2000) Development of hypothalamic neuronal thermosensitivity in birds during the perinatal period. *Journal of Thermal Biology* **25**: 119-123.

TZSCHENTKE, B. and BASTA, D. (2002) Early development of neuronal hypothalamic thermosensitivity in birds: influence of epigenetic temperature adaptation. *Comparative Biochemistry and Physiology Part A* **131**: 825-832.

TZSCHENTKE, B. and NICHELMANN, M. (1997): Influence of prenatal and postnatal acclimation on nervous and peripheral thermoregulation. *Annals of the New York Academy of Science* **813**: 87-94.

TZSCHENTKE, B. and NICHELMANN, M. (1999) Development of avian thermoregulatory system during the early postnatal period: development of the thermoregulatory set-point. *Ornis Fennica* **76**: 189-198.

TZSCHENTKE, B., SCHENDA, J. and PIERAU, FR.-K. (1994) Effects of cold adaptation on the modulatory action of TRH upon temperature sensitive and insensitive hypo-thalamic neurons. In: **ZEISBERGER, E., SCHÖNBAUM, E. and LOMAX, P.** (Eds) *Thermal balance in health and disease. Recent basic research and clinical progress:* pp. 67-72. Birkhäuser Verlag Basel.

TZSCHENTKE, B., BASTA, D., GOURINE, A.V. and GOURINE, V.N. (2000) Influence of bombesin on neuronal hypothalamic thermosensitivity during the early postnatal period in the Muscovy duck (*Cairina moschata*). *Regulatory Peptides* **88:** 33-39.

TZSCHENTKE, B., BASTA, D., JANKE, O. and MAIER, I. (2004): Characteristics of early development of body functions and epigenetic adaptation to the environment in poultry: focused on development of central nervous mechanisms. *Avian & Poultry Biology Reviews* **15:** 107 -118.

VIETH, E. (1989) Fitting piecewise linear regression functions to biological responses. *Journal Applied Physiology* **67:** 390-396.

WAIBEL, P.E. and MACLEOD, M.G. (1993) Associations between body weight and rectal temperature in turkey poults. *British Poultry Science* **34:** 417-423.

YAHAV, S. and MCMURTRY, J.P. (2001) Thermotolerance acquisition in broiler chickens by temperature conditioning early in life - the effect of timing and ambient temperature. *Poultry Science* **80:**1662-1666.

YAKIMOVA, K., SANN, H., SCHMID, H.A. AND PIERAU FR.-K. (1996) Effects of GABA agonists and antagonists on temperature sensitive neurons in the rat hypothalamus. *Journal of Physiology (London)* **494:** 217-230.

YAKIMOVA, K., SALLAGUNDALA, N. and TZSCHENTKE, B. (2005) Influence of baclofen on temperature sensitive neurons in chicken hypothalamus. *Methods and Findings in Experimental and Clinical Pharmacology* 27: 401-404.

YANG, H., ZHANG, C., ZHENG, H., XIONG, W., ZHOU, Z., XU, T. and DING, J.P. (2005) A simulation study on the Ca^{++} - independent but voltage-dependent exocytosis and endocytosis in dorsal root ganglion neurons. *European Biophysic Journal* **34:** 1007-1016.

ZHANG, C. and ZHOU, Z. (2002) Ca^{++}-independent but voltage-dependent secretion in mammalian dorsal root ganglion neurons. *Nature Neuroscience* **5:** 435-430.

Hypothalamic c-fos expression of temperature experienced chick embryos after acute heat exposure

O. JANKE AND B. TZSCHENTKE

Humboldt-University of Berlin, Institute of Biology, Perinatal Adaptation, Philippstraße 13, Haus 2, 10115 Berlin, Germany

The transcription factor and early immediate gene c-fos has been used as a stress marker to detect acute heat stress in chick embryos adapted to different incubation temperatures. On day 18 of incubation, three groups of embryos were subjected to warm, normal and cold temperatures until hatch. On the day of hatch, brains were dissected and sections of the anterior hypothalamus were assessed for c-fos immunohistochemistry. Heat stress evoked c-fos expression in all embryos but not in the unstressed controls. Trends which indicate differences between the three incubation groups were observed. These trends are in accordance with earlier results on the plasticity of the thermoregulatory system of avian embryos obtained from electrophysiological and metabolic investigations.

Keywords: c-fos, immunohistochemistry, epigenetic adaptation, chick embryo, heat stress

Introduction

The protein c-fos dimerises with the jun protein to form the transcription factor AP-1. AP-1 binds to different DNA-sites and thus influences the transcription rate over minutes to weeks. It is expressed within a short time after changes in the environment of the organism. Because of this, the c-fos gene is considered as an immediate early gene. It shows cell activation. In cellular models of synaptic plasticity, activation of immediate early genes mediates the rapid up-regulation of de novo RNA and protein synthesis (Lanahan and Wortey, 1998; Bock *et al.*, 2005). Compared to single cell recordings, the c-fos method can exhibit activity of numerous neurons at one time and thus demonstrate neuronal networks. In the nucleus of the cell, c-fos can be detected by immunohistochemical method. C-fos is often used as an unspecific stress marker. Thermal, osmotic and mental stress, mainly in rats, was detected using the c-fos method.

The polyclonal anti c-fos antibody sc-253 from Santa Cruz Biotechnology, Inc. is successfully used in mammalian and bird species. In the duck it was applied to detect c-fos expression in salt gland cells (Hildebrandt *et al.*, 1998) after osmotic stress. Harikai *et al.* (2003) have shown c-fos expression in mice preoptic area of the anterior hypothalamus (PO/AH) following heat stress. In the rat hypothalamus the above mentioned anti c-fos antibody was also used to detect c-fos expression following thermal load (Yoshida *et al.*, 2002; Tsay *et al.*, 1999).

E-mail: oliver.janke@rz.hu-berlin.de

Miyata *et al.* (1998) described that infant cold exposure changes c-fos expression to acute cold stimulation in adult hypothalamic brain regions. They exposed newborn rats until the age of 14 days for 2-4 h at 4°C every day. A cold stimulus of 10°C for 3 h on the 15th week evoked significantly lower neuronal c-fos expression when compared to the control, particularly in the hypothalamus. Another study (McKitrick, 2000) in adult rats under acute heat and cold exposure showed different intensity of c-fos expression in the hypothalamic area. Strong expression was found in the paraventricular hypothalamus, anterior central hypothalamus, medial-, ventromedial-, and dorsomedial hypothalamic area, supramammillary preoptic nucleus and in the ventrolateral preoptic nucleus during heat exposure.

In own preliminary investigations prior to this experiment cold incubated Muscovy duck and chicken embryos were challenged with acute exposure to cold on the last day of incubation against the normal incubated controls. The cold temperature experienced embryos showed a tendency to lower c-fos expression than the controls. Because of the high tolerance of bird embryos to temporary cooling, in the present experiments a heat stimulus was applied to evoke c-fos expression.

Earlier investigations of our working group related to neuronal plasticity of poultry species during early ontogeny have shown a clear influence of the incubation temperature during late incubation on the proportions of the thermosensitive neurons in the thermoregulatory centre in the preoptic area of the anterior hypothalamus (Tzschentke and Basta, 2002). During extra cellular single cell recordings it was demonstrated that cold incubation temperature led to a higher proportion of warm sensitive neurons and warm incubation was followed by an increase in the number of cold sensitive neurons. Within a few weeks after hatching these shifts disappeared probably due to the interference of acclimation to the actual postnatal environment (unpublished results). C-fos is expected to be a more sensible marker to show central nervous temperature adaptation for a longer period in birds as had been previously demonstrated in rats by Miyata *et al.* (1998).

The aim of the present study is to elucidate the detection of c-fos expression as a consequence of acute heat stress in avian embryos adapted to different incubation temperatures. The investigation is focused on the PO/AH. Results shall serve as a prerequisite for further investigations on the plasticity of the thermoregulatory system in embryos as well as adult birds.

Materials and methods

In the present study embryos of the layer chicken line white leghorn were investigated. Eggs were supplied by Lohmann Tierzucht GmbH, Cuxhaven, Germany. The embryos were incubated at 37.5°C and continuously turned in horizontal position. From day 18 of incubation onwards eggs were not turned. They were separated into three incubation groups. The controls were incubated at the regular 37.5°C. Two groups of embryos were exposed to altered incubation temperatures *viz* cold (34.5°C) and warm (38.5°C).

On the last day of incubation acute heat stress (42.5°C) for 90 min was applied to the embryos. Temperature and duration of the application were determined in previous experiments. Embryos showed a hyperaemia of the skin but survived. 5 embryos of all incubation groups (1 normal, 1 warm, 3 cold) did not receive acute heat stress.

Because incubation time varies at different incubation temperature experience from earlier experiments and candling of the eggs served to determine the last day of incubation.

Then the extracted embryos were anaesthetized and transcardial perfusion was performed. They were perfused with 50 ml phosphate-buffered saline (pH 7.4) and 150 ml of 4% paraformaldehyde in phosphate buffer (PB). After perfusion, brains were dissected and stored for an hour in the paraformaldehyde solution. Later they were transferred into 20% sucrose in PB and left overnight.

20 μm brain sections were made using a cryostate (Leica Microsystems, Wetzlar). The sections containing the hypothalamic region were attached to poly-L-Lysine coated micro slides. After washing in PB they were stored for one hour in a blocking solution containing 10% normal horse serum and 0.3% Triton X-100 (Sigma) in PB. Then the primary antibody (anti chicken c-fos, made in rabbit; Santa Cruz Biotechnology, Heidelberg) was incubated with the sections at 4°C for 48 h at a concentration of 1:1000. After washing with PB a biotinylated anti rabbit IgG (Axxora Deutschland GmbH, Grünberg) 1:200 in PB was bound to the previously formed complex during two hours incubation at room temperature. This incubation was followed by a second blocking step (0.3% H_2O_2 in methanol for 25 min) and washing in PB. By application of an ABC-solution (Vectastain Elite ABC Kit, Axxora Deutschland GmbH, Grünberg) horseradish peroxidase was attached to the immune complex. The latter served to bind diaminobenzidine tetrahydrochloride (DAB) 50 mg/100 ml (Sigma) in the presence of 0.0001% H_2O_2 in the next step. DAB evoke a brown staining of the c-fos positive cell nuclei of neurons. Later the sections were counter-stained using cresylviolet for visualization of the brain structures. Control sections were treated the same way without incubation with the primary antibody.

For analysis light microscopy and digital photography at a magnification of 50 fold (Zeiss Axioskop II, Zeiss AxioCam HRc) was used. C-fos positive neurons were counted in the PO/AH. Therefore a rectangle mask was placed between the *Comissura anterior* and the ventral margin of the brain around the third ventricle. Because of lacking stereotaxic data of the brain of the chick embryo, the width of the rectangle was set proportional to the PO/AH of the adult chicken at 990 μm for all embryos. Stereotaxic data of the adult chicken brain were taken from Kuenzel and Masson (1988). Means of three successive brain sections of each embryo were calculated for further analysis. Besides the PO/AH c-fos expression was examined in the whole sections.

The intensity of c-fos expression between embryos of the three incubation groups was compared with application of the non parametric Kruskal-Wallis-h-test because of the small number of investigations.

Results

The thermally stressed embryos showed the same vitality as the unstressed controls, but the colour of the skin had turned to light red.

In the control embryos without acute heat stress there were no neurons expressing c-fos in the PO/AH. A small expression was found in the hippocampus of one warm incubated embryo. Also the control sections without incubation with the primary antibody did not show c-fos expression.

In all differently incubated embryos c-fos was expressed after acute heat exposure. Table 1 shows the number of the c-fos positive neurons in the PO/AH of the three incubation groups. Besides the hypothalamic area minor and less regular c-fos expression was found in different brain areas as in the *Nucleus commissurae palli, Nucleus accumbens, Nucleus septalis lateralis, Hippocampus* and *Cortex piriformis*.

Table 1: Number of c-fos positive neurons after 90 min acute heat exposure in different incubated chick embryos on the last day of incubation. Positive neurons were counted in the preoptic area of the anterior hypothalamus.

Embryo number	Incubation time (d)	Incubation temperature D18-hatch (°C)	c-fos positive neurons	Group mean	Standard deviation
1	20	38.5	136	92	30
2	20	38.5	71		
3	20	38.5	80		
4	20	38.5	81		
5	21	34.5	81	128	55
6	21	34.5	62		
7	21	34.5	107		
8	23	34.5	59		
9	22	34.5	152		
10	22	34.5	147		
11	22	34.5	142		
12	22	34.5	226		
13	22	34.5	172		
14	21	37.5	101	180	89
15	21	37.5	129		
16	21	37.5	212		
17	21	37.5	136		
18	21	37.5	321		

The differences between the mean numbers of c-fos positive neurons in the PO/AH among the groups were not significant at $\alpha=0.05$ (Kruskal-Wallis-h-test). But two tendencies existed. Embryos without temperature experience showed the highest expression of c-fos on average (Fig. 1 C). Among the temperature experienced embryos the warm incubated embryos showed the smaller c-fos expression (Fig. 1 A) than the cold incubated embryos (Fig. 1 B).

Discussion

The results show, that acute heat stress on the last day of incubation led to a clear expression of c-fos in the PO/AH of all the investigated embryos when compared to the unstressed control. The absence of fos-staining in the unstressed control was also shown in the experiments of Hildebrandt et al. (1998) in the duck. The intensity and repeatability of the c-fos expression was much higher under heat stimulation than in the previously investigated preliminary results from the cold stimulated embryos. This difference is caused by the large tolerance of bird embryos to cold (Tazawa and Rahn, 1986).

The present trends as shown by the temperature-adapted chick embryos on the last day of incubation, have a smaller stress marker expression than inexperienced embryos and those with the opposite adaptation, are in accordance with results of electro-physiological and metabolic investigations of our group (Loh et al., 2004). Single cell recordings of hypothalamic neurons have shown a similar plasticity in cold and warm incubated duck and

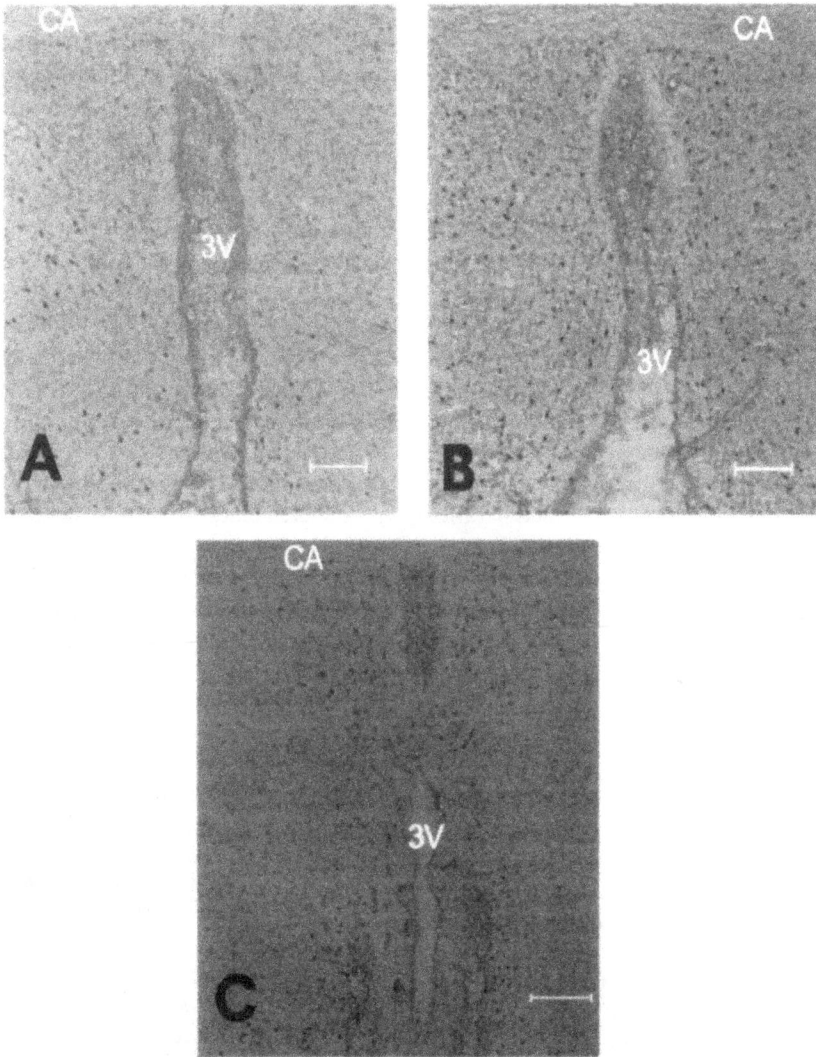

Figure 1. C-fos expression in the preoptic area of the anterior hypothalamus 90 min after acute heat exposure in a warm (A), cold (B) and normal (C) incubated chick embryo on the last day of incubation. C-fos positive nuclei of neurons are marked as dark dots. Scale bar: 100 μm. Abbrevations: CA, commissura anterior; 3V, third ventricle.

chicken embryos as in c-fos expression. Additionally, heat production was also higher in both groups by comparison with the control.

Embryos incubated at normal incubation temperature (37.5°C) were expected to have a medium heat stress between the cold and the warm incubated embryos after acute heat exposure. Possibly the fact that their thermoregulatory system was not challenged earlier, might have led to the highest expression value. In the course of embryonic development, stimulation of body functions due to changes in the environmental conditions induces as a rule first uncoordinated and immediate (proximate) non-adaptive reactions (Tzschentke *et al.*, 2004). It seems that during the early development of body functions it is not important for the organism that an adaptable reaction occurs but rather the fact that a reaction occurs anyway is important for adaptability during later life. At the end of embryonic or during the early postnatal period a qualitative change occurs in the reaction pattern of body function after environmental stimulation. The uncoordinated and/or immediate non-adaptive "training" reactions change into coordinated and/or adaptive reactions (Tzschentke *et al.*, 2004). Thus, a lack of activation might lead to less maturation of the respective system. The large variability of c-fos expression among individual embryos and cell activation in a number of brain regions outside the hypothalamus might be an expression of the immaturity of the brain functions of the embryo compared to that of an adult bird. During early ontogeny the initially existing high number of synapses in the brain is reduced to that which is getting regularly input and thus effectiveness of the system increases. Besides this, the majority of the c-fos positive nuclei outside the hypothalamus belong to the limbic system. It is known that this system is involved in the processing of emotions in man.

Postnatal investigations on heat stress in chickens using c-fos will give more insight. As such, they could also, if performed in differently incubated chickens, demonstrate the long term action of epigenetic temperature adaptation.

The findings in the hypothalamus of the adult rat under acute heat stress (McKitrick, 2000) are similar to those in chick embryos and should be compared to future results in adult chickens.

References

BOCK J., THODE, C., HANNEMANN O., BRAUN K. and DARLISON M.G. (2005b) Early socio-emotional experience induces expression of the immediate-early gene *ARC/ARG3.1* (activity-regulated cytoskeleton-associated protein/activity-regulated gene) in learning-relevant brain regions of the newborn chick. *Neuroscience* **133**: 625-633.

HARIKAI, N., TOMOGANE, K., SUGAWARA, T. and TASHIRO, S. (2003) Differences in hypothalamic Fos expression between two heat stress conditions in concious mice. *Brain Res. Bull.* **61**: 617-626.

HILDEBRANDT, J. P., GERSTBERGER, R. and SCHWARZ, M. (1998) *In vivo* and *in vitro* induction of c-fos in avian exocrine salt gland cells. *Am J Physiol.*, **275**, C951–957.

KUENZEL, W.J. and MASSON, M. (1988) A stereotaxic atlas of the brain of the chick (Gallus domesticus). *The John Hopkins University Press*, Baltimore: 62-63.

LANAHAN, A. and WORLEY, P. (1998) Immediate-early genes and synaptic function. *Neurobiology of Learning and Memory* **70**: 37-43.

LOH, B., MAIER, I., WINAR, A., JANKE, O. and TZSCHENTKE, B. (2004) Prenatal development of epigenetic adaptation processes in poultry: changes in metabolic and neuronal thermoregulatory mechanisms. *Avian Poult. Biol. Rev.* **15** (3/4): 119-128.

MCKITRICK, D.J. (2000) Expression of Fos in the hypothalamus of rats exposed to warm and cold temperature. *Brain Res. Bull.* **53**: 307-315.

MIYATA, S., ISHIYAMA, M., SHIBATA, M., NAKASHIMA, T. and KIYOHARA, T. (1998) Infant cold exposure changes Fos expression to acute cold stimulation in adult hypothalamic brain regions. *Neurosci Res.*, **31**: 219–225.

TAZAWA, H. and RAHN, H. (1986) Tolerance of chick embryos to low temperatures in reference to the heart rate. *Comp. Biochem. Physiol.* **85 A**: 531–534.

TZSCHENTKE, B., BASTA, D., JANKE, O. and MAIER, I. (2004) Characteristics of early development of body functions and epigenetic adaptation to the environment in poultry: focused on development of central nervous mechanisms. *Avian & Poultry Biology Reviews* **15**: 107-118.

TZSCHENTKE, B. and BASTA, D. (2002) Early development of neuronal hypothalamic thermosensitivity in birds: influence of epigenetic temperature adaptation. *Comp. Biochem. Physiol. A* **131**: 825-832.

YOSHIDA, K., MARUYAMA, M., HOSONO, T., NAGASHIMA, K., FUKUDA, Y., GERSTBERGER, R. and KANOSUE, K. (2002) Fos expression induced by warming the preoptic area in rats. Brain Res. **933**, 109-117.

Continuous, long-term observation of early development of chicken embryos *in ovo*

R. AKIYAMA*, A. KAMBARA, T. KOMORO, T. KATAOKA, H. YONETA, K. MORIYA[1] and H. TAZAWA

Department of Electrical and Electronic Engineering, Muroran Institute of Techonolgy, Muroran 050-8585; [1]Department of Electrical and Electronic Engineering, Hakodate National College of Technology, Hakodate 042-8501, Japan.

If fertile chicken eggs are incubated in an upright position, early development of embryos can be captured by a small camera through a window opened in the eggshell over an air cell. Accordingly, we attempted to develop a computer-aided video system which captured continuously the inside of the egg using a CCD camera installed directly onto the eggshell and to record early development of chicken embryos *in ovo*. The camera simultaneously functioned as a lid to close the window so that the embryos could survive several days or longer. The circular window 15 mm wide was opened through the eggshell over the air cell of fertile chicken eggs. The inner shell membrane was carefully peeled off by forceps to acquire a field of observation. An attachment comprising pieces of plastic cylinder 20 mm wide and sponge rubber 10 mm thick with an opening 15mm wide in the center was glued hermetically onto the eggshell surrounding the window. The opening was closed by an adhesive tape and the egg with the attachment was incubated in a still-air incubator warmed at 38°C. On Day 2 of incubation, a wide-scope CCD camera was inserted into the attachment. The inside of the egg was illuminated by eight green LEDs installed outside. The camera was focused occasionally on the embryo or the blood vessels. The video image was captured into the computer at 30 frames per sec and processed to present early development of the embryo, vitelline blood vessels, allantois and pulsating heart.

Keywords: CCD camera; early development; chicken embryo; vitelline circulatory system; pulsating heart; allantois

Introduction

The blastoderm of avian eggs is located on the top of the yolk which floats on the thick albumen. If the egg is incubated with the blunt end upwards, the embryo develops on the top of egg contents and early development of avian embryos *in ovo* can be seen through a window opened in the eggshell (Price and Fowler, 1940; Weiss and Andres, 1952; New, 1966). If the window opened over the air cell was covered adequately with a lid made from the broad-end half of the eggshell of another egg, or watch glass sealed on with wax, some of the embryos developed to the end of the incubation period (New, 1966). Development of embryos was observed when the

*Corresponding author: Ryuichi Akiyama
E-mail: rakiyama@mmm.muroran-it.ac.jp

lid was removed temporarily or through the lid, if still transparent. If the lid is replaced by a video camera connected to a computer, it is expected that early development of the embryo and extra-embryonic membranes is continuously recorded during a long period. The present study is designed to develop a computer-aided video system which captures continuously the inside of the egg using a small CCD camera inserted into the egg through an attachment glued onto the eggshell and records early development of chicken embryos *in ovo*.

Materials and methods

An attachment comprising pieces of a plastic cylinder and a sponge rubber board was used to install hermetically a camera inside an eggshell. The plastic cylinder 20 mm in diameter and 15 mm long was cut from a syringe. The sponge rubber board had a size 30 mm square and 10 mm thick. An opening 15 mm in diameter was cut in the center of the board and additionally a concentric circular track 20 mm wide and 5 mm deep was made around the opening. The rubber board was attached to an end of the plastic cylinder by inserting into the track and glued, providing the attachment for the camera. The opening through the rubber board of the attachment was closed hermetically by an adhesive tape until a camera was installed into it.

A circular window 15 mm in diameter was opened through the eggshell over the air cell of chicken eggs. The inner shell membrane was carefully peeled off by forceps to acquire the wide field of observation. The cylindrical end of the attachment was glued onto the eggshell surrounding the window. The egg installed with the attachment was incubated on upright position in a still-air incubator warmed at 38°C and relative humidity of about 55%. On Day 2 of incubation, the egg was transferred from the incubator to a bench and the adhesive tape was removed. Development of an embryo was checked through the hole. If the embryo developed, a wide-scope CCD camera (Type MTV-5366ND, Akizuki Denshi Tsusho Co. Ltd, Tokyo, capturing rate; 30 frames/sec, effective pixels; 352x240, lens diameter; 14.0 mm, color; 24-bit) was set through the hole. Heat originated from electricity in the camera was dissipated through a heat radiation board attached to the camera. The wire from the camera was hung from the ceiling of the incubator and connected to a computer. The inside of the egg was illuminated by eight green LEDs installed outside about 15 mm from the eggshell. The LEDs were placed at the height of the boundary between air cell and egg contents so as to surround the embryo. The green light with super-high intensity (6300-9500mcd, wave-length; 525 nm) was well absorbed by the blood vessels, and the camera was focused occasionally on the blood vessels or embryos. The video image was captured into the computer at 30 frames per sec with 352x240 pixels and MPEG-1 (Motion Picture Experts Group 1) format.

We attempted to abstract an image of vitelline blood vessels and the embryo using a technique similar to the companion report (Yoneta *et al.*, 2006). The individual color images captured by the CCD camera were differentiated into the three color components; that is, red, green and blue in 8-bit gradations (0-255), and divided into three pictures that were composed of one of three color elements. Because these three pictures were depicted by gradations of pixels of red, green or blue, the gradation range of pixels that composed preferentially the blood vessels and the embryo was assigned to individual color pictures. However, because the vitelline blood vessels and the embryo could not be differentiated from the background (i.e., yolk) in the blue picture, the blue picture was not used for the processing. The individual pixels comprising the original color images captured by the CCD camera were examined whether they satisfied the condition that the gradation of a given color pixel simultaneously stayed in the two gradation ranges (i.e., red and green elements). Then, the pixels which satisfied the condition were synthesized to abstract the blood vessels and the embryo. The area of abstracted images of the blood vessels and the embryo was expressed by the total number of pixels comprising them. The allantois was similarly abstracted and its area was also expressed by the number of pixels.

In order to present a time-course of heartbeat of pulsating heart, the heart was abstracted from the video image captured every 1/30 sec (i.e., 30 Hz) and its area was expressed by the number of pixels every 1/30 sec.

Results

Figure 1 presents an example of embryonic development every 10 hours from 70 hours of incubation. At 70 hours of incubation (panel A), the pulsating heart was dominantly recognized together with anterior and posterior vitelline veins and left and right omphalomesenteric arteries. The capillary network of the vitelline circulatory system (area vasculosa) was vague, but the sinus terminalis seen on right side and left upper corner separated the area vasculosa and the area vitellina. Ten hours later (panel B), the vitelline blood vessels grew widely over the screen, the heart was folded and the left omphalomesenteric vein was visible. The folding of the heart became pronounced with growth of the embryo, and the omphalomesenteric blood vessels (i.e., artery and vein) were clearly seen to parallel (panel C onward). The allantois was observed at 100 hours of incubation (panel D) and became large with time (panel E). Thereafter, the allantois grew further and the embryo moved left (upward on the screen) (panel F) with subsequent disappearance from the screen. Development of the allantois was presented in Figure 2. The allantois appeared at around 94 hours of incubation and its development was shown every 4 hours during 20-hour period.

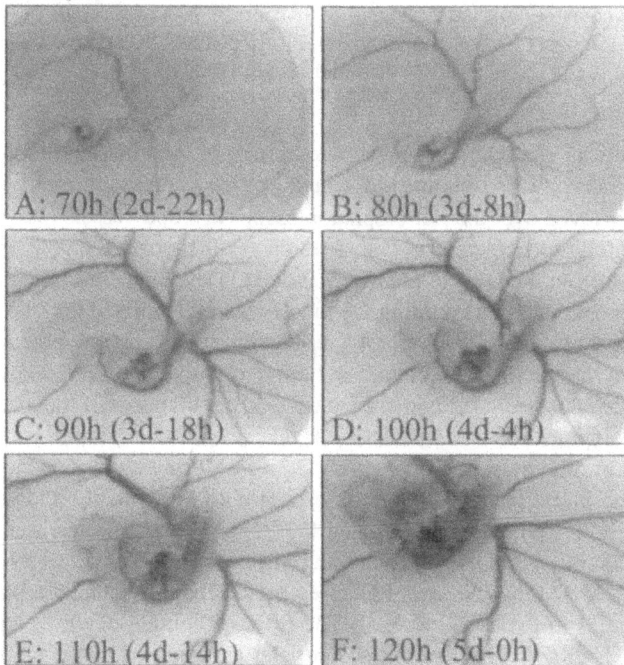

Figure 1. Embryonic development presented every 10 hours, taken from a video image captured during 50-hour period. A: embryo at 70 hours (2 days and 22 hours) of incubation, B: 80 hours (3 days and 8 hours), C: 90 hours (3days and 18 hours), D: 100 hours (4 days and 4 hours), E: 110 hours (4 days and 14 hours), and F: 120 hours (5 days).

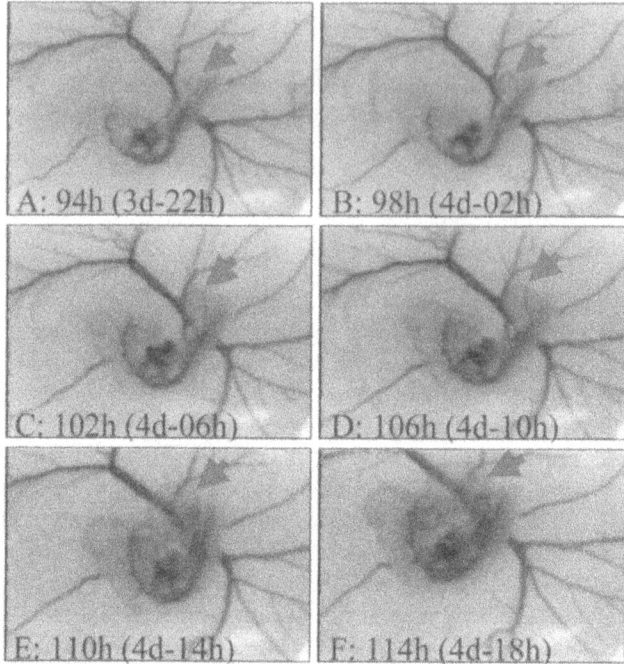

Figure 2. Development of the allantois of the embryo shown in Figure 1. Arrows indicate the allantois. A: allantois at 94 hours (3 days and 22 hours), B: 98 hours (4 days and 2 hours), C: 102 hours (4 days and 6 hours), D; 106 hours (4 days and 10 hours), E: 110 hours (4 days and 14 hours), the same as in panel E of Figure 1, and F: 114 hours (4 days and 18 hours).

Figure 3 presents an example of an abstracted frame of the vitelline blood vessels and the embryo. The embryo was 110 hours old, which was the same as in panel E of Figure 1. The allantois was observed over the left omphalomesenteric blood vessels (panel A). The blood vessels of the vitelline circulatory system developed widely over the yolk and the peripheral vessels were lost from the view of camera (panel B). Development of the blood vessels and the embryo was estimated by counting the pixels which comprised the blood vessels (panel B) and the embryo (panel C). Figure 4 presents the relative development of the embryo, blood vessels in the vitelline circulatory system and the allantois. Because the peripheral blood vessels were lost from the camera view already at 90 hours of incubation (C in Figure 1), relative development of blood vessels was decapitated after around 90 hours.

Figure 5 shows the pulsating heart every 2/30 sec during about one cardiac cycle. The heart ejected the blood through the aortic arches with very small residual volume of the ventricles at 0 msec (panel A) and dilated 2/30 sec (66 msec) later (panel B). Diastole continued for 200 msec and the ventricles were filled with the blood to form a round shape (panels C and D). During the next 133 msec period (panels E and F), blood was ejected from the ventricles to the aortic arches. The residual volume of the ventricles was small again 333 msec later (F). The area of the ventricles is expressed by the number of pixels for each frame. Figure 6 presents an example of the change in the ventricular area plotted every 1/30 sec for 5 sec in an embryo at 68 hours of incubation. The change in the area oscillates, indicating cardiac frequency; i.e., heart rate.

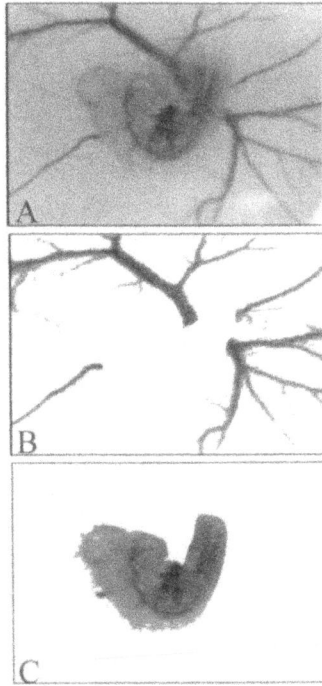

Figure 3. The blood vessels of the vitelline circulatory system and the embryo at 110 hours of incubation. A: video image of the blood vessels and the embryo. B: blood vessels. C: embryo.

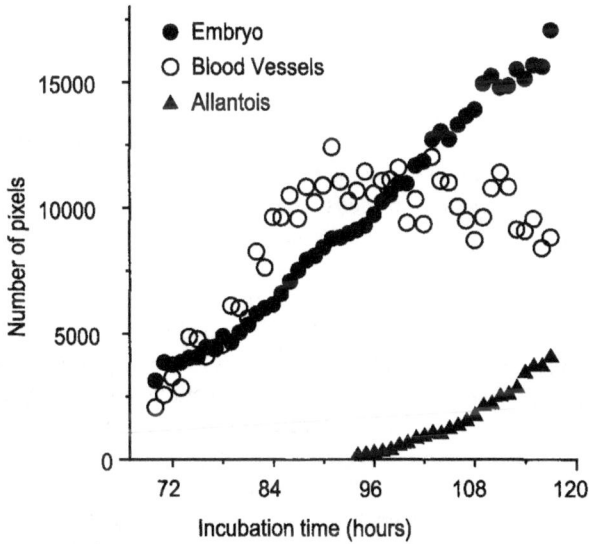

Figure 4. Relative development of the embryo (●), vitelline blood vessels (○) and allantois (▲), which was presented by the number of pixels, from 70 hours of incubation. The individual symbols are plotted every hour.

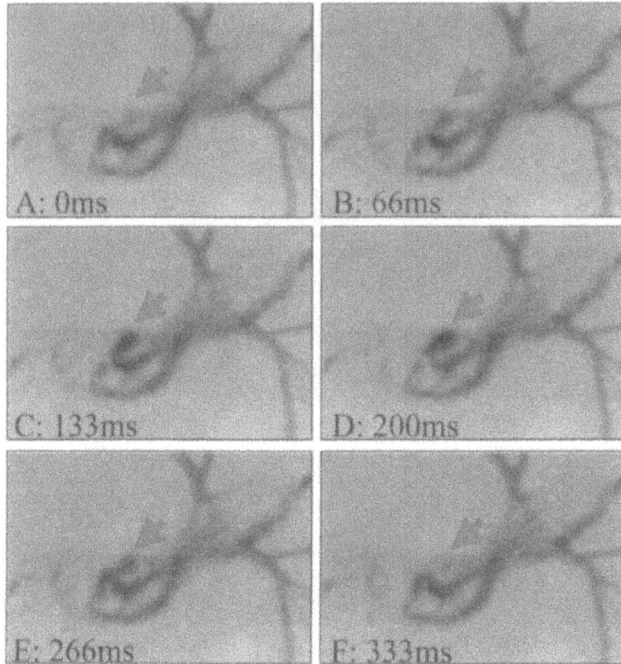

Figure 5. The pulsating heart presented every 2/30 sec (66 msec) at 80 hours of incubation. Arrow indicates the heart. The ventricular systole was at a final stage at 0 msec (panel A) and at 333 msec (panel F). The cardiac cycle was about 333 msec in this example. A: heart at time designated as 0 msec, B: 2/30 sec (66 msec), C: 4/30 sec (133 msec), D: 6/30 sec (200 msec), E: 8/30 sec (266 msec), and F: 10/30 sec (333 msec).

Discussion

The main goal of the present study was to ascertain the feasibility to determine continuously early development of the embryo living inside the eggshell. The goal was attained, but problems remained, and new feasibilities of application were expected. The inevitable problem was to peel off the inner shell membrane and its adverse effect was unknown. In the present experiment, the inner shell membrane was removed prior to incubation and the eggs with the camera attachment were incubated for more than 2 days. During incubation prior to experiment, the opening of the attachment was sealed hermetically to avoid water evaporation. On Day 2 (48-71 hours) of incubation, the eggs were examined by eye for development of the embryo through the opening and the camera was set through the attachment of the egg. The development of the embryo was confirmed by the CCD camera that displayed the embryo on a computer monitor. Because only eggs with live embryos were selected to develop a video analyzing system, a possible adverse effect of removal of the inner shell membrane on embryonic development remained to be investigated.

The eggs installed with the camera were returned to the incubator and the embryos were displayed on the monitor and captured up to 5 to 6 days of incubation until they moved out of the screen. The eggs were kept on the blunt end upwards while the embryos were videoed. Postures of all the embryos were not always in a fixed position. The embryo shown in

Figure 1 developed on the right side up. Some embryos developed on the left side up or on their face. The heart of the embryos growing on their side could be captured by the video system (Figure 5). Because the capturing rate is 30 frames per sec, the video system can be used to investigate early development of heart rate during the period of heart formation from a single ventricle with undivided atrium at about 65 hours to incomplete heart with four chambers at 120 hours of incubation (Romanoff, 1960). Judging from a correlation between chronological and structural age, the heart shown in Figure 5 comprised the single ventricle with incompletely separated right and left atria (Romanoff, 1960; Hamburger and Hamilton, 1951). The ventricle expanded and contracted once during about 10/30 sec (333 msec) in this example. Chronological consecutive changes in the ventricular area can be expressed by the number of pixels, yielding a pulsating signal of the heart (Figure 6). The cardiac pulsating signal formed an oscillatory pattern from which the heart rate can be approximated. Approximation of heart rate is about 160 beats per min for an embryo at 68 hours of incubation, which is close to the heart rate determined previously for 3-day-old embryos *in ovo* (Akiyama *et al.*, 1999). Improvement of the accuracy of heart rate determination is needed and depends on picture sharpness of the video image. The present system captured the video image with 352x240 pixels. Because the video image is captured by using a camera with 768x494 pixels, picture sharpness can be improved more than the present. In addition, wave-restoration of pulsating signal is expected to improve the accuracy of heart rate determination (Akiyama *et al.*, 1997), which remains to be studied. The improved picture sharpness is also required for determination of early development of the embryo, the allantois and blood vessels of the vitelline circulatory system.

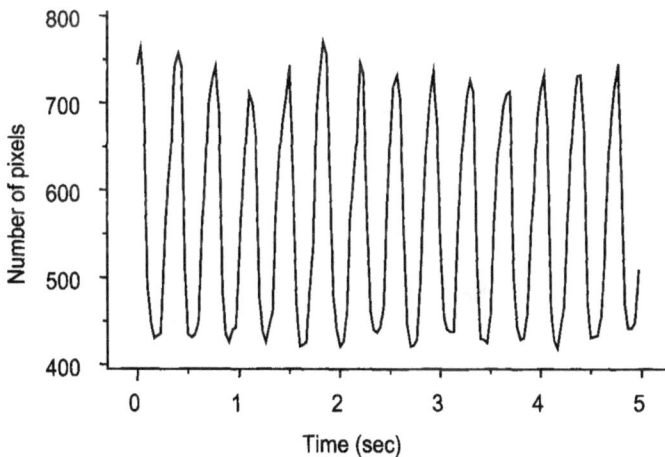

Figure 6. The cardiac pulsating signal of an embryo at 68 hours of incubation.

References

AKIYAMA, R., ONO, H., HÖCHEL, J., PEARSON, J.T. and TAZAWA, H. 1997. Non-invasive determination of instantaneous heart rate in developing avian embryos by means of acoustocardiogram. *Medical & Biological Engineering & Computing* **35**: 323-327.

AKIYAMA, R., MITSUBAYASHI, H., TAZAWA, H. and BURGGREN, W.W. 1999.

Heart rate responses to altered ambient oxygen in early (days 3-9) chick embryos in the intact egg. *Journal of Comparative Physiology* **B 169**:85-92.

HAMBURGER, V. and HAMILTON, H.L. 1951. A series of normal stages in the development of the chick embryo. *Journal of Morphology* **88**: 49-92.

NEW, D.A.T. 1966. *The Culture of Vertebrate Embryos.* Chapter 3 *The Chick (Gallus domesticus).* Academic Press, London, pp.47-98.

PRICE, J.W. and FOWLER, E.V. 1940. Egg shell cap method of incubating chick embryos. *Science* **91**: 271-272.

ROMANOFF, A.L. 1960. *The Avian Embryo.* Macmillan Publishing Co., New York.

WEISS, P. and ANDRES, G. 1952. The fate of embryonic cells (chick) disseminated by the vascular route. *Journal of Experimental Zoology* **121**: 449-187.

YONETA, H., AKIYAMA, R., NAKATA, W., MORIYA, K. and TAZAWA, H. 2006. Video analysis of body movements and their relation to the heart rate fluctuations in chicken hatchlings. In: *New Insights into Fundamental Physiology and Peri-natal Adaptation of Domestic Fowl.* (Ed. Yahav, S. and Tzschentke, B.), Nottingham University Press, Nottingham, pp. 57-68.

Effect of naked neck gene (*Na*) on embryonic mortality and embryonic metabolism in a heavy broiler breeder line

A. R. SHARIFI[1], P. HORST[2], W. SCHLOTE[2], H. SIMIANER[1] AND H. P. PIEPHO[3]

[1]Institute of Animal Breeding and Genetics, Göttingen, Germany. [2]Institute of Animal Science, Humboldt University, Berlin, Germany. [3]Institute for Crop Production and Grassland, University of Hohenheim, Germany.

Due to the *Na*-allele, embryonic mortality in the late hatching stage is significantly increased, the homozygous genotype being much more affected than the heterozygous embryo. The possible reason for the *Na* -gene induced embryonic mortality is still not clear. In this investigation it is shown that the phenomenon can neither be explained by the oxygen consumption, nor by the heat development or by the position of the naked neck embryo.

Keywords: Naked neck gene, embryonic mortality, oxygen uptake, broiler

Introduction

In a series of experiments a strong depressive effect of high ambient temperatures on growth and reproduction was verified. Depressions in important characteristics such as survival rate, laying activity and quality of hatching eggs as a result of the influence of high temperatures affect the profitability of the poultry industry in hot climates. In chicken breeding for tropical environments a useful strategy could be to create breeding stock carrying major genes for better heat tolerance. Particular major genes such as the sex-linked dwarf (*dw*), frizzled plumage (*F*), naked neck (*Na*) and sex-linked slow feathering (*K*) gene have been identified for their positive effects on heat tolerance (Horst, 1998). The *Na*-allele reduces feather coverage up to 40 % resulting in a relative heat tolerance. The higher productive adaptability of naked neck fowl under high ambient temperature, post embryonic vitality, and high carcass performance were demonstrated in many studies (Merat, 1990; Cahaner *et al.*, 1993; Yalcin *et al.*, 1997). But the commercial use of *Na*-allele is limited due to high embryonic mortality, which is caused by the *Na*-allele, where in particular the homozygous genotype is affected (Horst, 1982). The possible reason for the *Na*-allele induced embryonic mortality is still not clear. Embryonic malposition, perhaps due to fewer feathers on neck allowing the head to assume faulty positions (Merat, 1990), or metabolistic disorders (Decuypere, 2000, personal communication) have been held responsible for higher embryonic mortality in naked neck embryos. Therefore, in order to ascertain the source of embryonic death induced by the *Na*-allele special investigations are necessary. A series of experiments were conducted to determine the effect of naked neck gene on time of embryonic death, embryonic malposition, embryonic oxygen uptake and heat production.

E-mail: rsharif@gwdg.de

Materials and methods

STOCK AND HUSBANDRY

A heavy broiler sire line based on a commercial strain (Lohmann, meat) with genetic disposition for high growth performance, which was a carrier of the major gene for naked neck (*Na*), was used. 160 homozygous Naked neck (*NaNa*) and normal feathered (*nana*) hens were housed equally in separate cages under high and moderate temperatures, 30 °C and approximately 19 °C respectively, with relative humidity of 55-56 % from 20 to 72 weeks of age. During the rearing period and during the laying period the hens were kept under a restricted standard feeding programme (Lohmann, meat) with commercial diets. All birds were provided *ad libitum* access to water and had 12 hours of daily light.

For the analysis of impact of *Na*-allele of embryonic oxygen consumption and embryonic heat production eggs were obtained only from hens kept under temperate environmental temperature.

MEASUREMENTS

For the analysis of time of the embryonic mortality and malposition hatching eggs were obtained from 30th to 50th and from 54th to 72nd weeks of laying. Hens were artificially inseminated two times per week with 0.4 ml mixed and diluted semen of 24 sires of *nana* and *NaNa* genotype, which were kept under temperate climate conditions and restricted feeding programme (Lohmann management guides). Normal feathered (*nana*) females were crossed with homozygous naked neck (*NaNa*) and normal feathered males (*nana*) and naked neck (*NaNa*) females with naked neck (*NaNa*) and normal feathered (*nana*) males to obtain embryos of all three genotypes. The eggs were collected 3 times a day and incubated after a maximum storage time of 10 days at a storage temperature of 13 °C and relative humidity of 70 %. The eggs that were cracked or misshapen were removed.

Embryos that were dead at candling on 7th and 18th day of incubation as well as eggs that had not hatched were broken and examined macroscopically in order to estimate the time of embryonic death using methods of Hamburger und Hamilton (1951).

During incubation the chicken embryo moves itself to a normal position so that the long axis of the body is parallel to the long axis of the egg, with the head toward the blunt end. The head is turned to the right and lies under the right wing and the tip of the beak points towards the air cell. However, deviating from this normal position there are several unusual embryonic position which may hinder or prevent the embryo from hatching. These positions are termed malpositions (Romanoff, 1972).

The embryonic malposition of late-dead (16-21st incubation days) chicken embryos were classified into 7 different malpositions described by Landauer (1967), Romanoff (1972) and Anderson Braun (1988) as follows:

I   Head between thighs.

II   Head in small end of the egg (opposite air cell).

III   Head to left instead of right.

IV   Head in normal position but rotated with beak pointed away from air cell.

V   Feet over head.

VI   Beak (or head) over right wing.

VII Long axis of body differing from the long axis of the egg

From 14th to the 21st day of incubation the embryonic oxygen consumption was measured by putting hatchings eggs in four double wall incubation chambers with warm water circulating. The change of oxygen concentration in the incubation chamber was measured using a platinum-silver electrode system, which is described by Gauly *et al.* (2001) and Hiller (1994).

Embryonic heat production was measured from 16th to the 21st day by making one hole in the shell in the middle part of the eggs. Embryonic heat production was measured with a temperature-measuring sensor (Miniatur-Theroelement Ntc Siemens; accuracy 0.1 °C) on the egg membrane without damaging the membrane.

STATISTICAL ANALYSIS

Statistical analysis of embryonic mortality data was carrying out by application of a linear logistic model with a binary response variable, which is modelled as a binomial random variable (y_i). The dependent variable (y_i) can take the value 1 with a probability of success (survival up to certain time of incubation) π_i or the value 0 with probability of failure (embryonic death up to a certain time of incubation) $1-\pi_i$ for observation *i*. For relative frequencies the expected value (μ) equals the occurrence probability π_i. The logistic model uses a link function $g(\mu_i)$ linking the expected value to the linear predictors η_i. The logit link function is defined by

$$log\left[\frac{\pi_i}{1-\pi_i}\right] = \eta_i$$

Any change in η_i leads to a change in π_i. A basic form of generalised linear model can be described on the logit scale as $\eta_i = x_i\,\beta$, where x_i is the covariate vector for the ith observation and β is the regression coefficient vector. The linear predicor is linked to the original scale by $\eta_i = g_i(\mu_i)$, where $g(\mu)$ is the link function.

The embryonic mortality for different time period of embryonic development were considered as described by Beaumont *et al.* (1997):

Early embryonic mortality (EEM) occurs during the first week of incubation, middle embryonic mortality (MEM) occurs after 1st week of incubation and before transfer of the eggs into hatcher, and late embryonic mortality (LEM) occurs between 18th day of incubation to hatching. The data were then analysed with the GLIMMIX macro (Littell *et al.*, 1999) using the following generalized linear model

$$Logit\left(\frac{\pi_{rst}}{1-\pi_{rst}}\right) = \eta_{rst} = \varphi + \alpha_r + \beta_s + \tau_t + \alpha\beta_{rs} + \alpha\tau_{rt} + \beta\tau_{st} + \alpha\beta\tau_{rst}$$

where π_{rst} is the probability of EEM for a fertile egg or probability of MEM for a fertile egg that survives early embryonic development or probability of LEM of fertile egg that survives early and middle embryonic development; φ is the overall mean effect; α_r is the fixed effect of genotype of cock (levels: *NaNa, nana*); β_s is the fixed effect of genotype of hen (levels: *NaNa, nana*); τ_t is the fixed effect of environment (levels: warm, temperate); $\alpha\beta_{rs}$, $\alpha\tau_{rt}$, $\beta\tau_{st}$, $\alpha\beta\tau_{rst}$ are the fixed effects of interactions.

Since the brooding eggs, which are laid by the same hens, are correlated in their traits, or can be considered as repeated measurements, the modelling of serial correlations is necessary. A detailed description of modelling such serial correlations are given in Littell *et al.* (1999) and SAS Institute Inc., Cary, NC. (1997). The covariance matrix structure for repeated measurement for this data was fitted by using the Akaike information criterion. Model parameters were estimated by restricted maximum likelihood (REML). The Wald F-statistics were used to determine the significance of fixed effects on embryonic mortality (SS Type III - test).

Least square means were estimated on the logit scale and then back-transformed using the inverse link function $\pi = exp(x\beta) / [1+exp(x\beta)]$ to the original scale (probability) applying the LSMEANS statement. Significant differences between least square means were tested using a t-test procedure by inclusion of the PDIFF option in the LSMEANS statement. Standard errors of least square means were calculated as described by Littell *et al.* (1999).

A similar logistic linear model with the same explanatory variables was fitted for analysis of embryonic malposition considering malposition as binomial traits. In the first step of analysis all malpositioned types were combined into one class, thus data consisted of normally positioned and malpositioned embryos (normally positioned embryo=1, malpositioned embryo=0). In a second step of analysis the data consisted of normally positioned and the specific calcified malposition type (normal positioned embryo=1, specific calcified malpositioned embryo=0).

The oxygen uptake data was analysed by the MIXED procedure of SAS (SAS 8.01, 2000) with the following model:

$$y_{ijklmn} = \mu + E_i + H_j + D_k + I_l + C_m + EH_{ij} + ED_{ik} + HD_{jk} + EHD_{ijk} + bx_{ijklmn} + e_{ijklmn}$$

where y_{ijklmn} is the oxygen consumption, μ is the general mean, E_i is the fixed effect of genotype of embryo (levels: *NaNa, Nana, nana*), H_j is the fixed effect of hatch (levels: hatched, non-hatched), D_k is the fixed effect of day of incubation (levels: 14,...,19 days); I_l is the fixed effect of incubation chamber ($l = 1,...,4$); b is the linear regression coefficient of oxygen consumption on time between measurements within one day, EH_{ij}, ED_{ik}, HD_{jk}, and EHD_{ijk} are the fixed effects of interactions, C_m is the random effect of hens, and e_{ijklmn} is random error.

A similar linear model was used for the analysis of oxygen uptake at day of hatch with fixed effect of genotype of embryo (levels: *NaNa, Nana, nana*), fixed effect of hatch (levels: hatched, non-hatched), fixed effect of external pipping (levels: pipped, non-pipped), fixed effect of incubation chamber, linear regression coefficient of oxygen consumption on time between measurements in a day and interactions between main effects.

The statistical analysis of embryonic heat production data was carried out by using the following model:

$$y_{ijkl} = \mu + E_i + H_j + D_k + EH_{ij} + ED_{ik} + HD_{jk} + EHD_{ijk} + e_{ijkl}$$

where y_{ijkl} is the embryonic heat production, μ is the overall mean effect, E_i is the fixed effect of genotype of embryo (levels: *NaNa*, *nana*), H_j is the fixed effect of hatch (levels: hatched, non-hatched), D_k is the fixed effect of day of incubation (levels: 16,…,19), EH_{ij}, ED_{ik}, HD_{jk}, and EHD_{ijk} are fixed effect of interactions, and e_{ijkl} is random error.

Results and discussion

TIME OF EMBRYONIC MORTALITY

Table 1 summarizes the back transformed least square means and the significance of explanatory variable on embryonic mortality for different time periods of embryonic development. Neither the genotype of sire nor the genotype of hens as main factor have a significant effect of on EEM. However, a highly significant interaction was found between the genotype of hens and ambient temperature. The EEM of eggs laid from *nana* exposed to heat stress is clearly higher than of eggs of *nana* hens kept under temperate ambient temperature (22.6 % vs. 8.6 %). In contrast, there were no significant differences in EEM of eggs of naked neck birds kept under high and temperate environmental temperatures (13.8 % vs. 13.6 %). However, the EEM was somewhat higher for *NaNa* dams than for not heat stressed *nana* dams. A detailed analysis of data (result not shown) revealed that the 3rd day of incubation is the most critical time for embryonic survival of embryos obtained from heat stressed *nana* birds. A number of factors affecting embryonic liveability in this stage of embryonic development such as genetics, age of hens, time of oviposition, egg weight and quality as well as body temperature are described and summarized by Christensen (2001). Some of these factors are associated with stage of embryonic development at oviposition, which may lead to a suboptimal precondition for embryonic survival during storage and incubation. Exposing laying hens to heat stress results in a general retardation of laying activity (Petersen and Horst, 1978), reduction of clutch size, increase of time interval between two ovipotioned eggs, laying more eggs in the afternoon and impairment of egg and shell quality (Ahvar *et al.*, 1982; Sauveure and Picard, 1987), which might influence the developmental stage at oviposition and embryonic liveability (Christensen, 2001).

Lower shell thickness leads to higher egg shell conductions during storage and incubation and consequently to higher embryonic mortality (Romanoff, 1972; Roque and Soares, 1994). McDaniel *et al.* (1979, 1981) reported a significant deteriorative effect of low shell quality as measured by specific gravity on early embryonic liveability. Additionally, the bodies of heat-stressed hens have elevated body temperature (McDaniel, 1995), may lead to a higher production of abnormal embryos, although Ladjali *et al.* (1995) found no significant effect of heat stress on female gametogenesis or early cell division. They postulated that the decrease of hatchability under high ambient temperature is rather due to an altered metabolism during embryonic development that could result from a change in the biochemical composition of the egg suffering from heat stress.

For MEM (middle embryonic mortality), there was only a significant effect of environmental temperature on embryonic development, but no significant effect of dam or sire genotype was detected. There was no significant interaction between genotype of dam or sire and temperature.

Table 1. Least square means, standard errors and significance for different stages of embryonic death

Source of variation	Early embryonic mortality (1-7 d of incubation)		Middle embryonic mortality (8-17 d of incubation)		Late embryonic mortality (18-21 d of incubation)	
	LS-means[1]	Significance of difference	LS-means[1]	Significance of difference	LS-means[1]	Significance of difference
Ambient temperature (T)		≤0.0003		≤0.0001		≤0.5298
Warm	17.8±1.6[a]		6.3±0.7[a]		14.2±1.3[a]	
Temperate	10.9±1.0[b]		2.9±0.3[b]		13.2±0.9[a]	
Genotype of sire (S)		≤0.4082		≤0.8747		≤0.0001
NaNa	13.4±1.2[a]		4.2±0.5[a]		18.2±1.4[a]	
nana	14.6±1.1[a]		4.3±0.4[a]		10.2±0.8[b]	
Genotype of dam (D)		≤0.7631		≤0.1816		≤0.0001
NaNa	13.7±1.1[a]		4.8±0.3[a]		17.8±0.9[a]	
nana	14.3±1.5[a]		3.8±0.5[a]		10.4±1.1[b]	
S x D		≤0.4242		≤0.4038		≤0.0522
NaNa x NaNa	12.5±1.1[a]		5.1±0.4[a]		25.5±1.2[a]	
NaNa x nana	14.2±2.3[a]		3.5±0.8[a]		12.5±2.0[b]	
nana x NaNa	15.0±1.4[a]		4.6±0.5[a]		12.1±1.1[b]	
nana x nana	14.3±1.7[a]		4.1±0.6[a]		8.7±1.1[c]	
D x T		≤0.0004		≤0.5837		≤0.1428
NaNa x Warm	13.8±2.4[a]		6.8±0.5[a]		17.0±1.3[a]	
NaNa x Temperate	13.6±1.2[a]		3.4±0.4[a]		18.7±1.4[a]	
nana x Warm	22.6±1.9[b]		5.9±1.2[a]		11.8±2.2[a]	
nana x Temperate	8.6±1.7[c]		2.4±0.4[a]		9.2±1.0[a]	

[1] back transformed least square means.
Means for an effect within columns with no common subscripts differ significantly (p <.05)
Number of eggs set: N=9879; Number of examined eggs: N=2620.

For late embryonic survival neither the direct effect of temperature nor the interaction between temperature and other main effects were found to be significant, indicating that the deleterious effect of high temperature on hatchability occurs mainly during early embryonic development. However, the late embryonic liveability was highly affected by genotype of sire and dam and their interaction. The late embryonic mortality of heterozygous naked neck embryos was significantly higher than the one in normally feathered embryos, irrespectively from which sexes the *Na*-allele originated (8.7 % vs. 12.5 %, 12.1 %). The double dose effect on *NaNa* embryos caused a reduction of embryonic liveability of about 25.5 %. The comparison between the levels of late embryonic mortality in all of three genotype shows that the double dose effect of the *Na*-allele induces a higher embryonic mortality over the additive effect of both *Na* alleles. This result is in general agreement with those of Merat (1986, 1990), who report a diminishing effect of *Na*-allele on hatchability, without giving details about the quantity of late embryonic mortality caused by this gene. Gonzalez (1990) found no differences between all three genotypes of embryos in the proportion of pipped eggs of all non-hatched eggs indicating that *Na*-allele could not affect the mechanism involved in external pipping. Gonzalez (1990) assumed that the high rate of mortality of naked neck embryos may be attributable to an increasing effect of the *Na*-allele on egg weight and on egg shell thickness. A strong argument against this hypothesis is that the extent of embryonic mortality caused by the *Na*-allele is independent of the parent from which the *Na*-allele originated and so this is not subject to maternal effects. Regarding egg shell quality, a systematic influence of the *Na*-allele on egg shell quality could not be proven in many studies (Rauen, 1985; von Haaren-Kiso, 1991). However, comparative investigations on eggshell conduction during incubation may help to rule out this possibility.

EMBRYONIC MALPOSITION

The incidence of overall malposition for 1396 late dead embryos examined was 67.6 % (Table 2). The incidence of embryonic malposition II was high, while the incidences of malpositions I, V and VII were relatively low across all the explanatory variables. There were no significant effects of main factors as well as their interaction on frequency of malpositioned embryos. The incidence of a certain malposition was also not affected by main factors or their interactions. Data from the present studies indicate that traits associated with the presence of the *Na*-allele such as reduction of feathering on the neck (Merat, 1990) or an increase of egg weight does not result in a higher incidence of malposition irrespective of the genotype of the embryo or from which parent the gene originated. In comparison to *nana* embryos the incidence of normally positioned eggs in *NaNa* embryos is even higher.

EMBRYONIC METABOLISM

As already demonstrated above the presence of the *Na*-allele leads to a high late embryonic mortality. Embryonic mortality in chickens over the course of incubation is characterized by three periods; an early, a middle and a late phase. However, the most critical period for embryonic survival is the late phase with the peak of embryonic mortality at about day 19 of incubation (Romanoff, 1972; Jassim *et al.*, 1996). The late embryonic mortality coincides with the period in which the demand for oxygen increases significantly and with a series of physiological events such as initiation of pulmonary ventilation, external pipping and hatch from the shell.

The embryonic oxygen uptake during the first half of incubation is minimal; later it becomes exponential, and then it reaches a plateau for about three days just prior to the stage of internal pipping (Rahn, 1981). During the plateau stage the embryo experiences hypoxia and hypercapania due to the limitations of egg shell and membranes for gas exchange. The plateau stage is followed by the paranatal stage, which is characterized by penetration of the beak into the air cell (internal pipping) and the onset of pulmonary respiration, external pipping and hatching. The paranatal stage may also be affected by residual effects of the plateau in oxygen uptake.

Since *Na*-embryos show a higher mortality during embryonic development, one may suppose that the introduction of the naked neck gene leads to an increased demand for oxygen by the embryos during this phase, in turn leading to a reduction of embryonic liveability.

The influence of different genotypes on the oxygen demand of the embryos was investigated by measurement of oxygen consumption by different genotypes between the 14th and

Table 2. Least square means for embryonic position and the probability levels for analysis of variance[1]

Source of variation	Position of embryo (%)							
	Normal[2] vs. malpositioned	Malpositioned vs. normal[3]						
		I	II	III	IV	V	VI	VII
Ambient temperature (T)								
Warm	30.3	14.7	53.7	23.7	25.0	6.4	22.4	7.8
Temperate	29.6	14.0	54.7	17.2	31.0	4.2	18.8	11.6
Genotype of sire (S)								
NaNa	30.5	12.5	55.7	17.3	24.6	4.7	24.0	7.4
nana	29.3	16.4	52.5	23.6	31.4	6.0	17.4	12.2
Genotype of dam (D)								
NaNa	32.2	10.3	51.4	19.2	27.1	7.1	16.0	10.0
nana	27.7	19.2	56.9	21.4	28.8	4.4	25.8	9.0
S x D								
NaNa x NaNa	35.0	7.0	50.0	14.2	23.0	-	17.8	7.7
NaNa x nana	26.5	20.1	60.1	21.0	26.7	-	31.5	7.7
nana x NaNa	29.7	14.9	53.0	25.4	32.1	-	14.4	14.3
nana x nana	28.9	18.0	52.0	21.7	30.7	-	20.9	10.4
D x T								
NaNa x Warm	34.3	12.4	46.5	20.3	23.0	9.0	17.5	10.2
NaNa x Temperate	30.4	8.4	56.3	18.3	31.7	6.1	14.7	10.2
nana x Warm	26.5	17.3	60.7	27.6	27.1	4.5	28.2	6.0
nana x Temperate	28.8	22.3	52.0	16.3	30.3	2.8	23.6	13.2
Number and proportion of normal or malpositioned embryos of total examined embryos (1396)	453 32.4 %	59 4.2 %	468 33.5%	94 6.7 %	149 10.6 %	27 1.9 %	100 7.1 %	46 3.3 %

[1]The p-values of statistical tests are not given because there were no significant effects of main factors nor of their interaction on frequency of normal or specific malpositioned embryos.
[2]The transformed mean values showed the proportion of normal positioned embryos for different explanatory variables. All malpositioned types were combined as a non-normal positioned embryo, thus the data consisted of 453 normally positioned and 943 malpositioned embryos (normal vs. normal + all malpositioned embryos).
[3]The transformed mean values express the proportion of specific malpositioned embryos versus normal positioned embryos. The data for seven different statistical analysis consisted of normally positioned and a specific calcified malposition type. Accordingly, the total number of embryos in respective analysing group was: I=512, II=912, III=547, IV=602, V=480, VI=553, VII=499.

21st day of incubation. Table 3 shows the results of the analysis of the influencing factors on embryonic metabolism included in the model. As expected, the influence of the incubation day (14th to 19th) on the oxygen consumption of the embryos is highly significant. Embryonic oxygen consumption increases significantly between the 14th and 16th day of incubation (Fig. 1) and stays more or less constant - due to the limited diffusion of the eggshell - between the 16th and 19th day (plateau phase).

An effect of the embryo genotype on oxygen consumption could not be detected, neither as main effect nor in interaction with the factors hatchability and day of examination. There

Table 3. Significance of sources of variation in analysis of embryonic oxygen consumption and heat production

Source of variation	Oxygen consumption		Heat production
	14th d to 19th d of incubation p-value	Day of hatch p-value	p-value
Genotype of embryo (G)	≤0.9527	≤0.3843	≤0.0656
Hatchability (H)	≤0.0001	≤0.0001	≤0.0001
Day of incubation (D)	≤0.0001	≤0.0001	≤0.0001
Incubation chamber (C)	≤0.0001	≤0.0001	-
G x H	≤0.4060	≤0.6798	≤0.2410
G x D	≤0.9978	≤0.3151	≤0.0001
H x D	≤0.0138	≤0.2829	≤0.0557
G x H x D	≤0.5408	≤0.4365	≤0.4116
Rank	≤0.0276	≤0.1044	-

is no significant difference between the average (14th to 19th day of incubation) or daily oxygen consumption between the three genotypes investigated. An influence of the genotype could thus not be proven for non-hatchable and livable embryos.

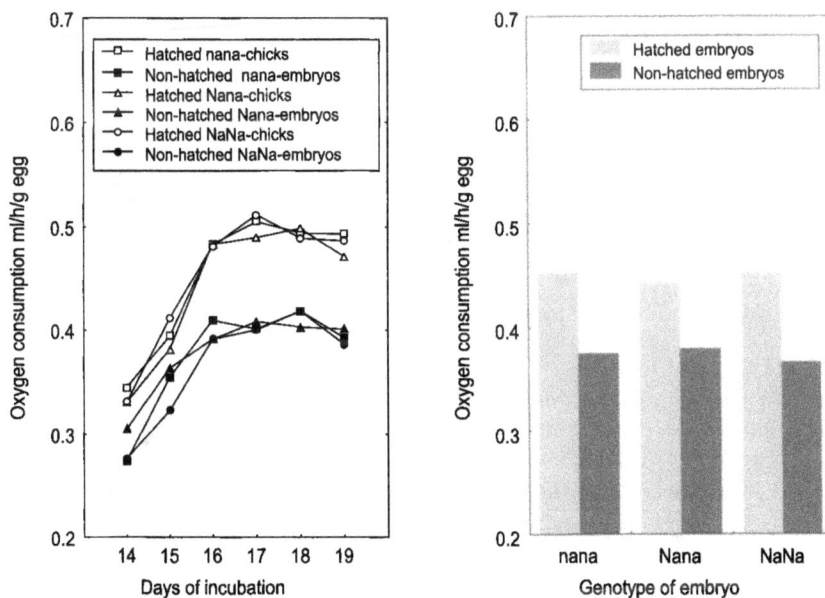

Figure 1. Oxygen consumption of hatchable and non-hatchable naked neck (*NaNa, Nana*) and normally feathered embryos during 14th - 19th day of incubation (left figure). The chart shows the average oxygen uptake of hatchable and non-hatchable naked neck and normally feathered embryos during 14th - 19th day of incubation (right figure).

The same holds for the individual days of incubation as no significant interactions could be shown to exist between the three main factors. On the other hand, the main factor hatchability is highly significant. The embryonic metabolism of eggs not bringing forth hatchable and livable chicks is significantly reduced as compared to eggs bringing forth livable chicks. Hatchable embryos consume an average of 20 % more oxygen over the whole range of incubation days (Figure 1).

The significance of the influencing factors on the embryonic oxygen consumption at hatching time (20th/21st day) is shown in Table 3. An influence of the genotype on the embryonic metabolism could not be shown for the day of hatch, supporting the findings for the period between the 16th and 19th day of incubation. External pipping abolishes the limiting effect of the eggshell on the diffusion of respiratory gases and thus leads to a significant increase in embryonic metabolism amounting to about 30 percent. But as there is no significant interaction between the main factors genotype and external pipping, there is no significantly increased rate of metabolism in naked neck embryos either, even when given free access to atmospheric oxygen. On the other hand, embryos that are not hatchable show a reduced metabolism on the day of hatch (Figure 2).

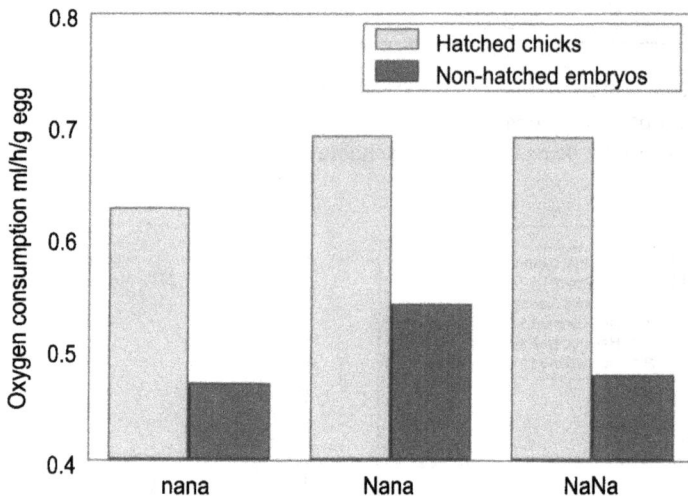

Figure 2. Least square means for embryonic oxygen uptake at day of hatch (ml/h/g egg)

Figure 3 shows that – as one would expect - the development of embryonic temperatures resembles the development of embryonic metabolism. Embryonic temperature initially increases between the 16th and 18th day of incubation.

There is no further increase up to the 20th day of incubation (plateau phase). After the 20th day of incubation there is another increase in temperature (post-plateau phase). This typical pattern is common with hatchable and non-hatchable embryos. But the hatchable embryos show a significantly higher temperature than the non-hatchable embryos (38.68 °C vs. 38.18 °C) over the whole period of examination. This is due to the higher rate of metabolism as already shown in regard to the embryonic consumption of oxygen. Table 3 shows a slightly significant influence of the embryonic genotype on the development of temperature. A comparison of the mean values shows this to be a result of the differences in temperature of non-hatchable embryos of differing genotypes (*NaNa, nana*) on the 16th day of incubation. Concerning the effect of the *Na*-allele on embryonic heat production during critical periods of late embryonic development, i.e., if a negative influence on embryonic liveability can be detected, the following conclusions can be drawn: On the one hand, the non-hatchable em-

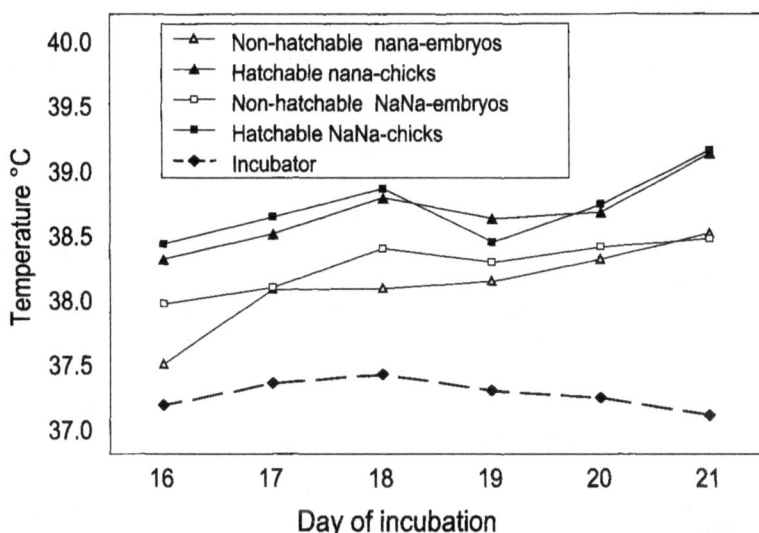

Figure 3. Embryonic heat production of hatchable and non-hatchable naked neck and normally feathered embryos during 16th – 21st day of incubation

bryos show a highly significantly reduced development of temperature as compared to the hatchable embryos, regardless of their embryonic genotype. On the other hand, there is no difference between the embryonic development of temperature between non-hatchable embryos of the naked neck type and the normally feathered ones. Embryonic hypothermia specifically caused by the naked neck gene cannot be proven and thus cannot be a possible cause for a higher embryonic mortality induced by the said gene.

During the late embryonic development anaerobic glycogenolysis is a major source of energy (Beattie, 1964; Freeman, 1965; Christensen and Donaldson, 1992). At oviposition, glycogen is hardly present in the egg. During embryonic development it is synthesised through gluconeogenesis and stored in the liver, heart and skeletal muscles. The glycogen content of the heart and liver is significantly reduced in the time between internal pipping and external pipping (Beattie, 1964; Freeman, 1965).

The thyroid hormones play an essential role in the synthesis of glycogen, in hepatic glycognesis and in the accumulation of glycogen in tissues with reduced gluconeogenesis (Christensen *et al.*, 1996). An artificial reduction of the permeability of the eggshell causing a higher grade of hypoxia leads to a reduction of glycogen-content in the liver, heart and skeletal muscles and a marked increase in late embryonic mortality (Beattie, 1964; Tazawa *et al.*, 1992). According to Beattie (1964), Christensen und Donaldson (1992) and Christensen *et al.* (1993; 1997) the heart glycogen stores at the beginning of the period of hypoxia and respectively a retarded growth of the embryonic heart during the period of hypoxia may limit the liveability of the embryo during this critical period of embryonic development.

In agreement with the results of experiments carried out by Hiller (1994), our own analysis of the oxygen metabolism has shown a marked reduction of metabolism in non-hatchable embryos (14th to 21st day of incubation) as opposed to hatched chicks (Fig. 1 and 2). Insofar, the oxygen metabolism before and during the period of hypoxia and after external pipping has to be taken into account when looking at the role carbohydrates play in the anaerobic metabolism. Looking at the embryonic metabolism of all three feather genotypes shows that there is no significant difference in oxygen consumption and thus in the oxygen metabolism of the three genotypes in regard to hatchable and non-livable embryos. Nor could a possible influence of the naked neck gene on the anaerobic metabolism be proven, because no differences in the embryonic temperature are detectable which is an indicator of the total embryonic metabolism. However, further investigations into the metabolic physiology of naked neck embryos are necessary to identify the effects of the naked neck gene on the embryonic metabolism in regard to the embryonic liveability.

The metabolism and the enzymatic activity of the M. complexus increase continually until the day of hatching, a process that uses glycogen as its main source of energy. Accordingly, the proportion of glycogen in the cells is very much reduced as the time of hatching approaches. During this time, the activity of enzymes involved in the glycogen-metabolism is relatively higher than the activity of oxygen related enzymes (Hagen, 1992).

Further research is needed to determine whether the naked neck gene has an effect on the physiology and the histologic structure or on the process of hypertrophia of the M. complexus which is important for the act of hatching. Possible influences on muscular, metabolic, hormone and enzymatic activity also need to be inspected, taking into consideration that the cutane district located above the muscle is affected by the reduced feathering caused by the naked neck gene.

Conclusions

Identifying the physiological basis of embryonic mortality caused by the naked neck gene is important for future production and breeding strategies. To further this, special research is necessary in fields such as:

- Investigation of embryonic movement and contraction of the head during external pipping using non-destructive techniques such as nuclear magnetic resonance imaging or 3-D sonography (Kuchida *et al.*, 1999; Narushin und Romanov, 2002).

- Comparative histometrical investigations of the M. complexus, possibly including the analysis of edema, hemorrhage and the number of necrotic cells which impair the hatchability (Rigdon *et al.*, 1968a; 1968b).

- Comparative investigations on weight development in embryonic heart, liver and hatching muscle during the plateau stage, during external pipping and hatch.

- Comparative histochemical investigations (glycogen content, metabolic waste products, enzyme and hormone activity) in embryonic heart, liver and hatching muscle during the plateau stage as well as during external pipping and hatch.

Acknowledgment

The authors would like to thanks Prof. V. Dzapo (Institute for Animal Breeding and Genetics, Justus-Liebig-University Giessen) and Prof. H. Tönhardt (Institute of Veterinary Physiology, Free University of Berlin) for their scientific and technical support.

References

AHVAR, F., PETERSON, J., HORST, P. and THEIN, H. (1982): Veränderungen der Eibeschaffenheit in der 1. Legeperiode unter dem Einfluss hoher Umwelttemperaturen. *Archiv für Geflügelkunde* **46**: 1-8.

ANDERSON BRAUN, A. F. (1988): Kunstbrut - Handbuch für Züchter. Hannover, M & H Schaper.

BEATTIE, J. (1964): The glycogen content of skeletal muscle, liver and heart in late chick embryos. *British Poultry Science* **5**: 285-293.

BEAUMONT, C., MILLET, N., LE BIHAN-DUVAL, E., KIPI, A. and DUPUY, V. (1997): Genetic parameters of survival to the different stages of embryonic death in laying hens. *Poultry Science* **76**: 1193-1196.

CAHANER, A., DEEB, N. and GUTMAN, M. (1993): Effect of the plumage-reducing naked neck (Na) gene on the performance of fast-growing broilers at normal and high ambient temperatures. *Poultry Science* **72**: 767-775.

CHRISTENSEN, V. L. and DONALDSON, W. E. (1992): Effect of oxygen and maternal dietary iodine on embryonic carbohydrate metabolism and hatchability of turkey eggs. *Poultry Science* **71**: 747-753.

CHRISTENSEN, V. L., DONSALDSON, F. and NESTOR, K. E. (1993): Embryonic viability and metabolism in turkey lines selected for egg production or growth. *Poultry Science* **72**: 829-838.

CHRISTENSEN, V. L., DONSALDSON, F. and NESTOR, K. E. (1997): Effects of an oxygen-enriched environment on the survival of turkey embryo between twenty-five and twenty-eight days of age. *Poultry Science* **76**: 1556-1562.

CHRISTENSEN, V. L., DONALDSON, W. E. and MCMURTRY, J. P. (1996): Physiological differences in late embryos from turkey breeders at different ages. *Poultry Science* **75**: 172-178.

CHRISTENSEN, V. L. (2001): Factors associated with early embryonic mortality. *World's Poultry Science Journal* **57**: 359 - 372.

FREEMAN, B. M. (1965): The importance of glycogen at the termination of the embryonic existence of *Gallus domesticus*. *Comparative Biochemistry and Physiology* **28**: 1169-1171.

GAULY, M., REINER, G., GRÖSCHL, M. and DZAPO, V. (2001): The effect of incubation condition on the embryonic respiratory and mitochondrial energy metabolism, the thyroid hormone status, and the daily gain in turkeys. *Archiv für Geflügelkunde* **65**(3): 97-105.

GONZALEZ, L., PEREZ, P., FRAGA, L. M., GUTIERREZ, O. and LERANDI, R. (1990): A note on the incubating performance of eggs from the inter-se cross of a heavy strain with naked neck heterozygous. *Cuban Journal of Agricultural Science* **24**: 323-326.

HAGEN, R. (1992): Zur Struktur und Bedeutung des Schlupfmuskels (M. complexus) beim Hühnerküken (Gallus domesticus): Dissertation, Veterinärmedizin. Berlin, Freie Universität Berlin.

HAMBURGER, V. and HAMILTON, H. L. (1951): A series of normal stages in the development of the chick embryo. *Journal of Morphology* **88**: 49-92.

HILLER, P. (1994): Untersuchung über den embryonalen und mitochondrialen

Energiestoffwechsel beim Legehuhn im Hinblick auf die Zuchtwertschätzung der Reproduktionsleistung. Gießen, Justus-Liebig-Universität Gießen.

HORST, P. (1982): Genetical perspective for poultry breeding on improved productive ability to tropical conditions. *2nd World Congress on Genetics Applied to Livestock Production*. Vol. **8**: 887-892.

HORST, P. (1998): Breeding possibilities in hot climates with special reference to tropical countries. 10th European Poultry Conference (p. 71-76), Jerusalem, Israel.

JASSIM, E. W., GROSSMAN, M., KOOPS, W. J. and LUYKX, R. A. J. (1996): Multiphasic analysis of embryonic mortality in chickens. *Poultry Science* **75**: 464-471.

KUCHIDA, K., FUKAYA, M., MIYOSHI, S., SUZUKI, M. and TSURUTA, S. (1999): Nondestructive predition method for yolk:albumen ratio in chicken by computer image analysis. *Poultry Science* **78**: 909-913.

LADJALI, K., TIXIER-BOICHARD, M., BORDAS, A. and MERAT, P. (1995): Cytogenetic study of early chicken embryos: effect of naked neck gene and high ambient temperatures. *Poultry Science* **74**: 903-909.

LANDAUER, W. (1967): The hatchability of chicken eggs as influenced by environment and heredity. Monograph I, Torres Agricultural Experiment Stations.

LITTELL, R. C., MILLIKEN, G. A., STOUP, W. W. and WOLFINGER, R. D. (1999): SAS system for mixed models. North Carolina, SAS Institute.

MCDANIEL, G. R., ROLLAND, D. A. and COLEMAN, M. A (1979): The effect of egg shell quality on hatchability and embryonic mortality. *Poultry Science* **58**: 10-13.

MCDANIEL, G. R., BRAKE, J. and ECKMAN, M. K. (1981): Factors affecting broiler breeder performance. 4. The interrelationship of some reproductive traits. *Poultry Science* **60**: 1792-1779.

MCDANIEL, C. D., BRAMWELL, R. K., WILLSON, J. L. and HOWARTH, B. (1995): Fertility of male and female broiler breeders following exposure to elevated ambient temperatures. *Poultry Science* **74**: 1029-1038.

MERAT, P. (1990): Pleiotropic and associated effects of major genes. In: *Poultry breeding and genetics*. Ed. R. D. Crawford. Amsterdam, Elsevier.

MERAT, P. (1986): Potential usefulness of the Na (naked neck) gene in poultry production. *World's Poultry Science Journal* **42**: 124-142.

NARUSHIN, V. G. and ROMANOV, M. N. (2002): Egg physical characteristics and hatchability. *Word's Poultry Science Journal* **58**: 297-303.

RAHN, H. (1981): Gas exchange of avian eggs with special reference of turkey eggs. *Poultry Science* **60**: 1971-1980.

RAUEN, H. W. (1985): Auswirkung des Gens für Befiederungsreduktion und Nackthalsigkeit (Na-Gen) auf das produktive Adaptationsvermögen von Legehennen an hohe Umgebungstemperaturen. Dissertation. Berlin, TU Berlin.

RIGDON, R. H., FERGUSON, T. M., TRAMMEL, J. L., COUCH, J. R. and GERMAN, H. L. (1968 a): Necrosis in the "pipping" muscle of the chick. *Poultry Science* **47**: 873-877.

RIGDON, R. H., BROWN, H. D., CHATTOPADHYAY, S. K. and PATEL, A. (1968 b): Pipping muscle of the chick. Clinical, pathological and enzymatic study. *Archive of Pathology* **85**: 208-212.

ROMANOFF, A. L. (1972): Pathogenesis of the avian egg. New York, Wiley-Interscience.

ROQUE, L. AND SOARES, M. C. (1994): Effects of eggshell quality and broiler breeder age on hatchability. *Poultry Science* **73**: 1838-1845.

PETERSEN, J. and HORST, P. (1978): Hohe Umwelttemperatur als auslösender Faktor von Genotyp x Umwelt-Interaktion beim Legehuhn. *Archiv für Geflügelkunde* **42**: 173-178.

SAS INSTITUTE INC., CARY, NC. (1997): SAS/STAT Software: change and enhancements through release 6.12. SAS Institut.

SAUVEUR, B. and PICARD, M. (1987): Environmental effects on egg quality. Current problems and recent advances (Poult. Sci. Symposium Series). R. G. Wells and C. G. Belyavin. London, Butterworths. 20.

TAZAWA, H., HASHIMOTO, Y., NAKAZAW, S. and WHITTOW, G. C. (1992): Metabolic responses of chicken embryos and hatchlings to altered O_2 environments. *Respiration Physiology* **88**: 37-50.

VON HAAREN-KISO, A. (1991): Bedeutung des Gens für Lockenfiedrigkeit (F) unter besonderer Berücksichtigung der Kombination mit anderen Majorgenen für das produktive Adaptationsvermögen von Legehennen an hohe Umwelttemperaturen. Berlin, Technische Universität Berlin.

YALCIN, S., TESTIK, A., OZKAN, S., SETTAR, P., CELEN, F. and CAHANER, A. (1997): Performance of naked neck and normal broilers in hot, warm and temperate climates. *Poultry Science* **76**: 930-937.

Compensatory renal hypertrophy in chick embryos with experimentally induced unilateral renal agenesis

*Z.ZEMANOVÁ, M.J. MURPHY[1]

Institute of Physiology, Academy of Sciences of the Czech Republic, Vídeňská 1083, 142 20 Prague 4, Czech Republic; [1] Department of Natural Sciences, College of Agriculture and Technology, State University of New York, Cobleskill, NY, 12043 USA

We used an animal model, the chick embryo, with experimentally induced unilateral renal agenesis, URA, to analyse the consequences of a prenatal failure of one kidney on the subsequent development of the embryo and on the contralateral kidney. We evaluated the development of the embryo (survival, body weight, gross malformations) as well as development of the remaining kidney (organ weight, nephron diameter) and of other abdominal organs following the URA induction at stage HH 10 – 12. We found that more than 30 % of operated embryos survived until day 18, though no hatching occurred due to eventration of thoraco-abdominal organs. Body weight of embryos with URA was comparable to that of controls between days 7 and 12. The remaining mesonephros, however, appeared to grow faster than mesonephroi of control embryos, but the difference between their wet weights was significant only on day 8, compared to the doubled weight of remaining mesonephros. The relative weights of the remaining mesonephros were significantly higher not only on day 8 (7.8±1.9 vs 6.1±0.9 mg/g), but also on days 9 and 10, when the difference also amounted 20 %. Only at day 12, weight of the remaining mesonephros was lower than that of control kidneys. The faster growth of remaining mesonephros correlated with a significantly enlarged diameter of proximal tubules: at day 7 it increased from 51.8±8.8 μm by 24 % and at day 8 the diameter of 58.7±11.7 μm enlarged by 18 %. Gonads and adrenals in embryos with URA developed macroscopically normally. The liver of 9 – 10-day embryos with URA consistently exhibited white plaques, which were likely to be uric acid deposits.

This model exhibits two phases in development of embryos with URA: first, giving evidence of compensatory renal hypertrophy, and second, manifesting signs of chronic renal failure.

Key words: chick embryo - mesonephros - unilateral renal agenesis - compensatory growth - relative kidney weight - proximal tubular diameter - gonads - liver

Introduction

Unilateral renal agenesis (URA) is a genetically or epigenetically conditioned serious renal disorder occurring in humans as well as in animals. In the human population it occurs rela-

*Corresponding author E-mail: zemanova@biomed.cas.cz

tively frequently - on average of 1 in 1000 births (see e.g. Wilson and Baird 1985, Yuksel and Batukan, 2004). A detailed analysis of URA could be helpful in delineating the adverse effects of the missing kidney to the embryo and in strategies to prevent the collapse of the remaining kidney function. URA appears in humans as a solitary malformation (Hill, *et al.*, 2000), or as a part of various complex syndromes (Stewart *et al.*, 1999, McCallum *et al.*, 2001). It is caused by a disorder in the development of the Wolffian duct or of the ureteric duct derived from it. It is also caused by invalid inductive interactions between the ducts and nephrogenic blastema or by a failure of the proper interconnection between nephron anlages and the collecting system. It seems, therefore, likely to be multifactorial and usually includes one or more genetic components; consequently, the disorder frequently includes other components of the urogenital system in addition to the kidney. Similar disorders appear also in laboratory animals, where URA occurs either "spontaneously" (i.e. without visible cause), or is induced by genetic manipulations, as in the case of inbred mouse strain FUBI with failure of ureteric bud invasion (Kamba *et al.*, 2001), or by intrauterine surgical unilateral nephrectomy – commonly performed in fetal lambs (Moore *et al.*, 1987) or in fetal rabbits (Abellan *et al.*, 1997). An especially elegant animal model of URA induction is accomplished by surgical intervention before mesonephric development in non-mammalian embryos developing outside the uterus (Bishop-Calamé, 1966). It consists in an interruption of the caudal migration of the Wolffian duct and results in the absence of both kidneys (i.e. the mesonephros and the metanephros) on the operated side, resulting in URA. It must be performed very early, just at the beginning of organogenesis of the kidney. In this respect it quite closely mimics spontaneous renal agenesis, in contrast to unilateral nephrectomy in the fetus. The early and persistent developmental effects of URA on the entire embryo are, thus, striking as repeatedly demonstrated (Wenz *et al.*, 1992, Zemanová and Murphy, 1998).

The absence of one kidney, as observed in most cases studied after a nephrectomy, induces a series of events leading to postnatal compensatory growth of the remaining kidney (Argyris *et al.*, 1969, Pfaller *et al.*, 1998). Compensatory growth in these cases includes the enlargement of the contralateral kidney, an increase of glomerular size and of filtration area, and an increase of the tubular mass. The growth was faster in younger individuals compared to adults (Argyris *et al.*, 1969, Celsi and Jakobsson, 1986). In contrast to these studies we are interested in the succession of the compensatory reactions in prenatal periods. In human embryos, growth of the remaining kidney was measured sonographically as an increase of the kidney length (Steele *et al.*, 1993, Hill *et al.*, 2000). Compensatory growth was detectable *in utero* as early as 22 weeks of gestation. In animal models, similar responses of the remaining kidney were recorded. For example, seventy two hours after left uninephrectomy in fetal lambs at midgestation age the intervention induced an increased weight and renal cortical content of RNA and DNA (Moore *et al.*, 1979). A significant increase in renal weight and an increase of the glomerular area, indicating compensatory growth, was also observed in fetal rabbits three days after uninephrectomy (Abellan *et al.*, 1997).

In this paper we present our observations performed on the chick embryo. This animal model has the distinct advantage of allowing an examination of URA from the earliest stages of development. We used the above mentioned procedure of Bishop-Calamé (1966). The procedure allows extensive observation of the remaining contralateral mesonephros and the subsequent metanephros, if desired (Wenz, *et al.*, 1992). Our aim was to evaluate development of the remaining kidney and the impact of missing kidney on development of the chick embryo during first half of the incubation. The study of the model disorder of URA in early embryos is likely to bring important knowledge of critical points in their development and the strategies for the protection of the remaining kidney from the consequences of the over-load.

Material and methods

INCUBATION OF EMBRYOS AND SURGERY EVOKING URA

Fertilized eggs of outbred Grey Leghorn breed (Kolec farm, Institute of Molecular Genetics, Academy of Sciences, Czech Republic) were incubated at a temperature of $37.3 \pm 0.3\ ^\circ C$ and relative humidity 55 – 65 % for 36-42 h. Access to the embryos was enabled by a rectangular opening of the egg shell. The operation bringing about the URA was performed after Bishop-Calamé (1966) by interrupting the path of the caudal migration of the meso-nephric (Wolffian) duct. This was accomplished by the small incision made with a glass knife in the left lateral body wall below the level of $10^{th} – 16^{th}$ somite. To prevent healing of the interrupted body wall a small piece of sterile eggshell membrane was inserted into the incision. The drilled window above the embryo was covered with a parafilm square sealed to the window by a warmed instrument. Control embryos were left without operation. The embryos were operated at stage HH 9 – 12, and their survival was daily monitored.

Collection and examination of embryos

The embryos – both operated and controls – were incubated until day 7, 8, 9, 10, or 12. Then they were collected using decapitation, weighed, checked for external malformations and dissected. The state of liver, gonads, adrenals, and mesonephric kidneys was recorded and kidneys were then removed and weighed. Renal weight was evaluated as a wet weight of actually developed mesonephroi (two or one in embryos with URA). For comparison of growth of mesonephroi in embryos with URA and controls there were still counted double renal weight in embryos with URA, and a relative weight, related to the body weight.

Measurement of the tubular diameter

External tubular diameter was measured in embryos with a surgically exposed mesonephros on day 5 (Friebová-Zemanová *et al.*, 1982), placed into the so-called "chick incubation bath" (Zemanová *et al.*, 1993). During the recording of a transepithelial potential difference in mesonephric tubules of 7- and 8-day embryos, the proximal tubular dimensions were measured *in vivo*.

Statistics

For statistical evaluation of data the Student-Neuman-Keuls multiple range test was used. Evaluation of significance of differences in growth of mesonephros weight was performed using the two-way analysis of variance and covariance with two grouping factors: "experimental group" (URA and controls) and "age" (days 7, 8, 9, 10, 12) with one covariant factor, body weight. The age groups were also evaluated separately on days 7 – 10 and on day 12. Differences in proportions of embryos with lesions of liver and with cystically dilated tu-bules in mesonephros were statistically evaluated with the use of the Fisher exact test.

Results

EFFECTIVENESS OF SURGERY AND SURVIVAL OF EMBRYOS AFTER THE EX-
PERIMENTAL INTERVENTION (TABLE 1)

**Table 1. Survival of the chick embryos after the surgery inducing an absence of the
left-sided mesonephros (URA) in dependence of the stage of intervention**

| Stage at surgery | Number of operated embryos | Survival of embryos with URA until | | | | |
		day 5 %	day 8 %	day 9 %	day 12 %	day 18 %
HH 10	38	84	45	39	27	-
HH 11	30	80	67	67	37	37
HH 12	17	76	47	47	47	-

Surgical intervention to arrest the caudal growth of the left-sided mesonephric (Wolffian)
duct and to prevent the induction of the left mesonephros was typically successful in about
90% of embryos operated on stages HH 10 - 12. URA was also successfully obtained in
embryos operated later (HH 13 – 14), but in such cases, a fragment of the cranial portion of
the left mesonephros developed on the operated side, resulting in incomplete URA.

Early mortality (Table 1) before incubation day 5 was minimal in embryos that un-
derwent surgery at stages HH 10, 11 or 12 (16%, 20%, and 24% of operated embryos,
respectively). The mortality of embryos did not depend on the stage of embryos at surgery.
It increased by 15 – 40 % between days 8 and 9. This was partially caused by gross, observable malformations of the body and/or extraembryonic membranes. These included a delay
in allantoic outgrowth and origin of the chorioallantois, the organ important for exchange of
respiratory gases. The later decrease of surviving embryos was minimal and embryos with
the experimentally induced URA could, surprisingly, survive until day 20. However, hatching did not occur due to the growth retardation connected with permanent defects of the
body wall.

INCIDENCE OF GROSS MALFORMATIONS (TABLE 2)

**Table 2. Incidence of malformations recorded in 10-day embryos (% of examined em-
bryos)**

Group	Caudal regression, rhachischisis	Microphtalmia	Ectopia cordis, eventration of viscera	Normal
URA	30	17	70	22
Controls	0	0	0	100

Operated embryos that died early (incubation days 2 and 3) exhibited caudal blisters filled
with a clear fluid or blood (though probably non-communicating with forming blood vessels).

The left-side URA surgery induced a retardation of development of the embryos and the
malrotation of the embryos; this included head and neck rotated left side down and the body/

trunk lying on its ventral or right side instead of the left side position. This heterotaxis was found in all 5-day operated embryos. Conversely, control embryos rotated completely to their left side. This malrotation in response to URA persisted until embryonic day 7 in many of successfully operated embryos, but on day 9 it was found only rarely.

Gross malformations of the axial structures - caudal regression, rhachischisis, and malformations of the head – microphtalmia and exencephaly were recorded from day 5.

However, the main developmental anomaly was a large opening of the body cavity resulting in the irrecoverable eventration of thoracic (*ectopia cordis*) and abdominal organs of various extent. This condition was associated with a defect in retraction of the remainder of yolk into the abdomen.

GENERAL GROWTH (TABLE 3)

Table 3. Growth of embryos with URA and growth of the mesonephroi

Age	URA BW (mg)	MsW (mg)	Double MsW (mg)	Rel.Ms$_U$W[†] (mg/g)	Controls BW (mg)	MsW (mg)	Rel.Ms$_C$W[‡] (mg/g)
(d)	(n)	(n)	(n)		(n)	(n)	
7	588±84[*)	1.9±0.4	3.8±0.8	6.54±0.96	628±86	4.1±0.6	6.66±0.82
	(17)	(17)	(17)		(14)	(14)	
8	972±142	3.7±0.8	7.5±1.7	7.78±1.90**	1019±113	6.1±1.0*	6.11±0.91**
	(14)	(14)	(14)		(18)	(18)	
9	1546±133	4.9±0.8	9.7±2.08*	6.32±1.36*	1471±156	6.9±1.2*	4.82±0.91*
	(12)	(12)	(12)		(7)	(7)	
10	1772±277	4.7±0.9	9.33±1.9	5.36±1.28**	1882±291	7.9±2.2	4.15±0.91**
	(22)	(20)	(20)		(10)	(10)	
12	3834±267	5.8±2.4	11.6±5.4*	3.04±1.44	3952±376	16.6±2.1*	4.21±0.50
	(5)	(6)	(5)		(13)	(13)	

Explanations:

*) ... mean ± SD
* $p<0.05$, ** $p<0.01$... statistical significance of differences between URA and control groups
† ... relative weight of doubled weights of single mesonephros in embryos with URA
‡ ... relative weight of paired mesonephroi in controls

Growth of the embryos was evaluated using body wet weight in all experimental and control groups at 7d, 8d, 9d, 10d and 12 d. The difference in body weight between embryos with URA and controls was not significant. But the differences in BW between particular age groups were highly significant ($p<0.001$).

APPEARANCE AND GROWTH OF THE MESONEPHROI (TABLE 3, FIGURE 1)

The ventral aspect of the shape of the mesonephroi in the dissected embryos could be orientatively used for an evaluation of the progress of renal development. The model shapes of the mesonephroi recorded in a ventro-dorsal projection at stages 5 to 10 days (Friebová,

1975) differed so much that they enabled us to distinguish in remaining mesonephroi their accelerated development. This was recorded from day 8 (a shift in progress by half a day) and was apparent also on day 9.

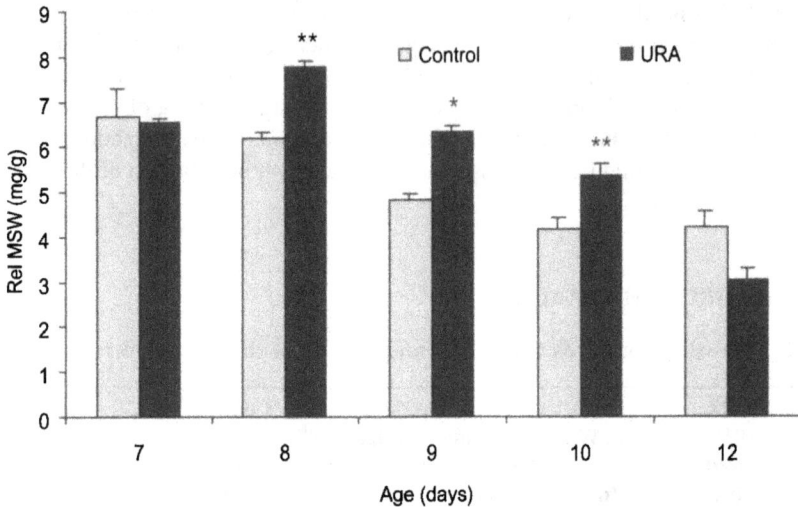

Figure 1. Relative weight of mesonephros (Rel MSW) in embryos with URA and controls. Rel MSW (mg/g body weight) was equal in embryos with URA (black columns) and in controls (gray columns) only at day 7. At days 8 – 10 the difference between experimental and control embryos was statistically significant (* = p<0.05, ** = p<0.01). The lower value of Rel MSW in experimental embryos at day 12 was not significantly different from the control one.

Growth of mesonephroi was characterized by wet weight changes. The weights of the remaining mesonephros are presented in Table 3 first as the weight of a single mesonephros, and second as a double of the actual weight to enable a comparison of growth of the mesonephros of embryos with URA with both kidneys of controls.

The two-way analysis of variance and covariance (with covariant factor body weight) of differences between renal weight in embryos with URA and controls revealed that the difference between groups "URA" and "controls" was non-significant (p=0.644), contrary to the highly significant differences between ages (p<0.01). It appeared when days 7 – 12 were analysed together. Omitting day 12 from the two-way analysis resulted in an increase of significance of renal weight differences both between groups and ages. This signified that growth of the mesonephroi progressed in the group with URA and in controls differently beginning at 12 days of age. Doubled weights of single mesonephros in embryos with URA were significantly higher than weights of paired mesonephroi in controls only on day 9 (9.7±2.08 vs. 6.9±1.2; p<0.05), but lower than in controls on day 12 (11.6±5.4 vs. 16.6±2.1; p<0.05). The relative weights of remaining mesonephroi (again doubled to be comparable with relative weights of paired mesonephroi in controls) were significantly higher in URA than in control group on days 8 and 10 (7.78±1.90 vs. 6.11±0.91, and 5.36±1.28 vs. 4.15±0.91, respectively, p<0.01) and also on day 9 (6.32±1.36 vs. 4.82±0.91, p<0.05). A one-way analysis of variance and covariance testing the differences between ages separately in experimental groups revealed that the influence of body weight on the weight of the mesonephros was significant in controls, but not in embryos with URA. Differences between average mesonephros weights of day 7 to day 10 were significant only in embryos with URA (in day 7 versus days 8 and 9 - p<0.01, in day 7 versus day 10 - p<0.05).

NEPHRON GROWTH (TABLE 4)

Table 4. Size of the external diameter of the proximal tubules of the mesonephroi measured *in vivo*

Age (d)	Controls (μm)	n	URA (μm)	n
7	51.8 ± 8.8*	15	64.3 ± 16.9*	13
8	58.7 ± 11.7	33	69.1 ± 29.1*	8

* ... mean ± SD

Measurements were performed on tubules of 7-day and 8-day embryos (Table 4). The external PT diameter was significantly larger (p<0.05) in embryos with URA compared to controls in both age groups (by 24 % on embryonic day 7 and by 18 % on day 8).

The enlargement of the tubular lumen in some cases ranged to double that of normal. This effect comes under the category of a pathological dilation commonly referred to as a cystic dilation of the tubule (CDT).

INCIDENCE OF CYSTIC DILATIONS OF TUBULES OF MESONEPHROS

Cystic dilations of tubules, CDT, were observed in 22 % of the remaining mesonephros of 67 embryos with URA and in 10 % of kidneys of 39 control embryos. The difference is not statistically significant. Both values occur below the border of 30 %, which was declared as the limit for a teratologically significant incidence of malformations. The incidence of "clouded" tubules (instead of tubules with clear transparent walls) was in our material very scarce (once in 106 embryos).

OBSERVATIONS OF ADJACENT ABDOMINAL ORGANS

The gonads as well as *adrenals* developed in embryos with URA as paired organs and were, in gross aspect, quite normal. The left gonad was always present in a position slightly shifted to the medial line from its usual location due to the missing left mesonephros. Adrenals were both situated in their regular location.

The *liver* in embryos with URA grew normally, but in its parenchyma often appeared white plaques occupying one or several places at the margin of the organ. We found them significantly more frequently in embryos with URA (in 33 cases representing 50 % of the total) than in controls (0 – 12 %) at age 9 – 10 days. The difference was statistically significant (p=0.0001) when tested with the two-sided Fisher exact test.

Discussion

Unilateral renal agenesis is a developmental disorder, a malformation of the urinary system, which can be compatible with relatively normal life, when the defect appears without additional abnormalities. This is evident in the high incidence of reported birth defects of URA in human populations. Such a reduction of excretory function, combined with compensatory growth of the remaining kidney, probably represents a prenatal adaptation. It cannot be interpreted as a simple renal failure, as in mammalian embryos the excretory function is mostly performed by mother's excretory organs.

We do not know, however, what is the total incidence of patients with URA in human populations, because embryos dying early are lacking for analysis and diagnosis. In our "chick model of URA" we observed ca. 20 % mortality in URA embryos during the period of embryogenesis; it is unknown whether this is directly due to URA or to some other consequence of the surgical intervention. It seems highly probable that our observed 70 % incidence of thoracal and abdominal eventrations may be a consequence of the paraaxial incision necessary for interrupting growth of the Wolffian duct. The insertion of a piece of sterile eggshell membrane into the incision, especially a little larger piece (used to ensure a successful arrest of the Wolffian duct growth), may impair development of the body wall and of other axial structures and cause the eventration of organs as well as caudal regression. The resulting gross malformations are therefore different in chick embryos compared to naturally-occurring URA in humans, where malformations of the urogenital system alone are more common.

Avian embryos are, in some respects, more severely compromised by the adverse effects arising from the absence of a single kidney and the subsequent low amount of urine produced. Avian embryonic urine forms allantoic fluid necessary for the expansion of the allantois/chorioallantois; this is a trivial concern for placental mammals. The chick embryos cannot survive with the poor development of the chorioallantoic membrane since, between days 7 and 8, it takes over the respiratory function of the richly blood-supplied vitelline membrane. The critical decrease of urine production stops growth of the chorioallantois and it clearly can lead to the death of experimental embryos.

Surprisingly, body growth of experimental embryos with URA and of controls did not differ in spite of the developing growth retardation in older experimental embryos. The difference in body weights could be masked, however, by more or less pronounced edema appearing in embryos with URA. We will clearly need additional data on dry weights of embryos in future experiments.

Growth of the remaining kidney in our study, however, corresponded well to that described in mammals. The compensatory growth of the contralateral kidney in adult mice, for example, exceeded the control values by 10-20% at 24 h after uninephrectomy (Argyris *et al.*, 1969), while the increase in the kidney weight in adult rats was 33 % 2-3 weeks after uninephrectomy (Katz and Epstein, 1967). In chick embryos with URA the weight of remaining mesonephros exceeded weights of mesonephros in controls at age groups 8, 9, and 10 days by 23 – 41 %. On day 7, however, there was found no difference, and on day 12 the double weight of the remaining kidney in embryos with URA amounted already only 70 % of the control kidneys. The excessive growth of the remaining mesonephros was thus apparent only on days 8 – 10. We hypothesize, that at a younger stage, on day 7, the hypertrophy was probably not yet fully expressed. By day 12, on the contrary, the slowing down of growth of the mesonephros could likely manifest the onset of renal failure as has been reported in earlier studies (Wenz *et al.*, 1992). The embryonic kidney in birds excretes uric acid as an end-product of nitrogen metabolism and its failure to perform this function could intoxicate the embryos with URA. Between days 12 and 18 only 37 % of operated embryos survived in our study, whereas between days 8 and 12 most of them survived. It seems evident that some new negative factors start to work by day 12 of embryonic life.

Earlier studies reported an increase of glomerular size and of available filtration area in animal models with one kidney after uninephrectomy (Celsi *et al.*, 1989), and an increase of the tubular mass with the proximal tubule growth in length (Celsi *et al.*, 1986). Our data are similarly assessing the proximal tubular diameter, where we found an increase of 18 – 24 % in the remaining mesonephros. Neither true cystic dilations of the tubules, one of the signs of the tubule overload in embryos, nor so-called clouded tubules, indicative of increased concentrations of uric acid, however, appeared rarely in our experimental material. In contrast, a marked pathology was found in liver, where obvious white plaques were routinely observed at the margin of the lobules. This we considered as a possible mark of the renal failure characterized by an inability to excrete uric acid, produced not only by the mesonephros but also by the liver.

Conclusions

Our model for unilateral renal agenesis in the chick embryo demonstrates several characteristic features of early embryonic response to a renal insufficiency. These include: a progressive increase in weight of the remaining mesonephros, indicating marked compensatory renal growth, and a subsequent slowing down of the growth, indicating a start of the renal failure to excrete uric acid. These findings are strikingly similar to mammalian embryos lacking one kidney. The chick model of URA, however, is the only model that currently allows a detailed embryonic analysis as it is easily accessible during all stages of development, including those before and during organogenesis.

Acknowledgments

Authors are grateful to Mrs Miluše Strnadová for her expert technical assistance, and to Mrs Alena Dĕdicová for statistical analyses. This work was supported by research grant from the Grant Agency of the Czech Republic No. 304/04/0972

References

ABELLAN, M. C., CHEHADE, A., GRIGNON, Y., GALLOY, M. A., FABRE, B. and SCHMITT, M. (1997) Compensatory renal growth post fetal nephrectomy in the rabbit. *European Journal of Pediatric Surgery* 7: 282-285.

ARGYRIS, T. S., TRIMBLE, M. E. and JANICKI, R. (1969) Control of induced kidney growth. *Compensatory renal hypertrophy.* NOWINSKI, W. W. and GOSS, R. J. Academic Press, NY, London.

BISHOP-CALAMÉ, S. (1966) Étude expérimentale de l'organogenesc du systeme urogénitale de l'embryon de poulet. *Archives de Anatomie et Microscopique Morphologie Experimentale* 55: 217-309.

CELSI, G. and JAKOBSSON, B. (1986) Influence of age on compensatory renal growth in rats. *Pediatric Research.* **20**: 347-350.

CELSI, G., LARSSON, L. and APERIA, A. (1986) Proximal tubular reabsorption and Na-K-ATPase activity in remnant kidney of young rats. *American Journal of Physiology (Renal Fluid Electrolyte Physiology 20)* **251**: F588-F593.

CELSI, G., LARSSON, L., SERI, I., SAVIN, V. and APERIA, A. (1989) Glomerular adaptation in uninephrectomized young rats. *Pediatric Nephrology.* **3**: 280-285.

FRIEBOVÁ, Z. (1975) Formation of the chick mesonephros. 1. General outline of development. *Folia Morphol.* **23**: 19-28.

FRIEBOVÁ-ZEMANOVÁ, Z., KUBÁT, M. and CAPEK, K. (1982) Experimental approaches to the study of kidney function in chick embryos. *The Kidney during Development. Morphology and Function.* Ed. SPITZER, A., Masson Publishing, New York USA.

HILL, L. M., NOWAK, A., HARTLE, R. and TUSH, B. (2000). Fetal compensatory renal hypertrophy with a unilateral functioning kidney. *Ultrasound Obstetrics and Gynecology* **15**: 191-193.

KAMBA, T., HIGASHI, S., KAMOTO, T., SHISA, H., YAMADA, Y., OGAWA, O. and HIAI, H. (2001) Failure of ureteric bud invasion: a new model of renal agenesis in mice. *American Journal of Pathology* **159**: 2347-53.

KATZ, A. I. and EPSTEIN, F. H. (1967) Relation of glomerular filtration rate and sodium reabsorption to kidney size in compensatory renal hypertrophy. *The Yale Journal of Biology and Medicine* **40**: 222-230.

MCCALLUM, T., MILUNSKY, J., MUNARRIZ, R., CARSON, R., SADEGHI-NEJAD, H. and OATES, R. (2001) Unilateral renal agenesis associated with congenital bilateral absence of the vas deferens: phenotypic findings and genetic considerations. *Human Reproduction* **16**: 282-8.

MOORE, D., TUDEHOPE, D., LEWIS, B. and MASEL, J. (1987) Familial renal abnormalities associated with the oligohydramnios tetrad secondary to renal agenesis and dysgenesis. *Australian Paediatric Journal* **23**: 137-41.

MOORE, E. S., DELEON, L. B., L.S., W., B.J., M. and OCAMPO, M. (1979) Compensatory renal hypertrophy in fetal lambs. *Pediatric Research* **13**: 1125-1128.

PFALLER, W., SEPPI, T., OHNO, A., GIEBISCH, G. and BECK, F. X. (1998) Quantitative morphology of renal cortical structures during compensatory hypertrophy. *Experimental Nephrology.* **6**: 308-319.

STEELE, B. T., MCGRATH, F. P. and GLAZEBROOK, K. N. (1993) Prenatal compensatory renal growth: Documentation with US. *Radiology* **189**: 733-735.

STEWART, T. L., IRONS, M. B., COWAN, J. M. and BIANCHI, D. W. (1999) Increased incidence of renal anomalies in patients with chromosome 22q11 microdeletion. *Teratology* **59**: 20-22.

WENZ, J. R., PECK, M. P. and MURPHY, M. J. (1992) Unilateral renal agenesis in chick embryos: a model for chronic renal insufficiency. *International Journal of Developmental Biology* **36**: 445-450.

WILSON, R. D. and BAIRD, P. A. (1985). Renal agenesis in British Columbia. *American Journal of Medical Genetics* **21**: 153-69.

YUKSEL, A. and BATUKAN, C. (2004). Sonographic findings of fetuses with an empty renal fossa and normal amniotic fluid volume. *Fetal Diagnotics and Therapy* **19**: 525-32.

ZEMANOVÁ, Z. and MURPHY, M. J. (1998). Unilateral renal agenesis in the chick embryo: Evaluation of the contralateral mesonephric kidney. (Abstract).7th International Workshop on Developmental Nephrology "Early renal development – a key to the understanding of adult diseases?". Ed.CELSI G. Stockholm, Sept 9 - 11.

ZEMANOVÁ, Z., UJEC, E. and KUBÁT, M. (1993). A new technique for studying potential changes in nephrons of chick embryo kidneys *in vivo*. *Journal of Developmental Physiology* **19**: 37-41.

The prenatal development of the avian claw - a model for the formation of function-related modified skin. A preliminary report

RUTH MARIA HIRSCHBERG[1], HERMAN BRAGULLA[2] AND KLAUS-DIETER BUDRAS[1]

[1] Institute of Veterinary Anatomy, Faculty of Veterinary Medicine, Freie Universität Berlin, Koserstraße 20, D-14195 Berlin, Germany; [2] currently: Department of Biological Sciences, Louisiana State University, 202 Life Science Building, Baton Rouge, LA 70803-1715, USA

The avian claw, i.e., the modified skin covering the tip of the distal phalanx, was developed for protection of the phalanx and, species-specifically, as an organ for grasping, digging, climbing, and defence. Accordingly, this digital end organ shows particular function-related structural modifications in all layers (epidermis, dermis and subcutis) of the skin, resulting in the formation of the different species-specific shapes of the avian claws.

Whereas the epidermis produces the strongly cornified claw capsule, the connective tissue layers of the claw - dermis and subcutis - provide specific anchorage and force transmitting structures besides the neuro-vascular supply for the digital end organ.

The prenatal development of these specific integumental structures therefore serves as an excellent model for studying the functional adaptation of the skin, as well as for a comparative morphological approach to study the structure and function-related modifications in the development and in the phylogeny of the digital end organs in general.

The prenatal development of the avian claw was examined using light microscopic and scanning electron microscopic techniques on the feet of different developmental stages of cockatiel (*Nymphicus hollandicus*), duck (*Anas platyrhynchos* forma *domestica*) and chicken (*Gallus gallus* forma *domestica*).

The present preliminary report focuses on the development of the general structures of the avian digital end organ, i.e. the epidermis, the dermis, and the subcutis, and especially on the formation of the pododermal vasculature.

The epidermis of the developing avian claw produces different generations of cornifying cells, thus forming a deciduous claw capsule ensheathing the permanent horn capsule, produced in the last trimester before hatching. The mesenchymal dermis underneath the epidermis produces collagen fibres spanning from the third phalanx to the

*Corresponding author: Dr. Ruth M. Hirschberg
E-mail: hirschberg.ruth@vetmed.fu-berlin.de

dermo-epidermal interface. A 'typical' subcutaneous layer was found in the distal reticulate scales, but not in the different regions of the claw proper. Additionally, a separate desmal ossification of the tip of the distal phalanx was detected. The development of specific vascular plexus in the connective tissue compounds of the claw anlage allowed distinguishing of the prospective dermal and subcutaneous layers even in the early, otherwise less differentiated stages of claw development. Thus, ossification of subcutaneous tissue in some areas of the developing claw was detected - homologue to the apical part of the distal phalanx.

Keywords: Avian claw; digital end organ; dermo-epidermal interface; pododermal vasculature; subcutaneous ossification

Introduction

A literature review reveals that claws play an increasing role in fowl husbandry - e.g., claw reduction or claw removal in commercial poultry industry (Compton *et al.*, 1981; Vanskike and Adams, 1983; Hargis *et al.*, 1989; Frankenhuis *et al.* 1991; Honaker and Ruszler, 2004), whereas comparatively little is known on the structure and functions of the avian claw (Moser, 1906; Boas, 1931; Lucas and Stettenheim, 1972; Spearman, 1985).

The claw represents the typical form of the 'digital end organ'[1] in birds, and is comprised of the modified skin covering the tip of the distal phalanx. In the course of phylogeny, the avian claw was developed for protection of the distal phalanx and, species-specifically, as an organ for grasping, digging, foraging, climbing, and defence.

Accordingly, the digital end organ displays particular function-related structural modifications, resulting in the formation of the different species-specific shapes of avian claws: e.g., the falculae - curved and sharp pointed claws - of raptor birds, compared to the blunter and more compact claws of ratites; or the more infrequent development of claws at the tip of the wings (Moser, 1906); e.g. in hoatzin chicks (*Opisthocomus hoazin*; Anonymous, 2006), enabling it to climb and grip tree branches.

Because they represent specific modifications of the digital skin, the end organs are always comprised of all three layer of the skin, i.e. epidermis, dermis and subcutis (Zietzschmann, 1918) with respective modifications.

Whereas the *epidermis* of the claw is characterised by distinct keratinisation and enhanced cornification compared to the non-modified skin and thus produces the strongly cornified claw capsule (see fig. 1A), the connective tissue layers of the claw, i.e. the *dermis* (see fig. 1B) and the *subcutis*, provide specific anchorage and force transmitting structures besides neuro-vascular supply for the digital end organ, thus enabling vaso-neuro-hormonal coordination of all integumental layers (Homberger, 2000; Hirschberg *et al.*, 2001a; Bragulla and Hirschberg, 2003). These functions are correlated to the development of specifically shaped surface modifications of the *dermo-epidermal interface* - i.e. a prominent papillary or lamellar body - with corresponding *angioarchitecture*, as described in detail for the equine

---

[1] Digital end organ: This general morphological term describes all different shapes of homologue end organs in birds, reptiles and mammals, such as claws, nails and hooves; defined in the Nomina Anatomica Veterinaria (NAV, 2005).

hoof (Bragulla, 2003), the bovine hoof (Hirschberg *et al.*, 2001a) or the human nail (Sangiorgi *et al.*, 2004). The specialised connective tissue cushions of the digital *subcutis* are functionally related to shock absorption and occur either integrated into the digital end organ (e.g. within the periople, coronet and bulb of the equine and bovine hoof) or situated adjacent to the digital end organ in the form of digital pads (e.g. in the canine claw; described by von Süsskind-Schwendi, 2005). According to Moser (1906), no subcutis is developed in the claw proper of birds but is distinctly detectable within the adjacent digital pad and unguicular scale.

Because in these specific modifications each integumental layer may vary within the different regions of all digital end organs (claws, hooves, nails) and in order to achieve homology and homonomy when comparing for example avian and mammalian claw development (Hirschberg *et al.*, 2001b), the end organ is subdivided into five different regions (or segments): the periople, the coronet, the wall proper, the sole and the bulb (Zietzschmann, 1918; NAV, 2005). As already mentioned above, the bulb may either be integrated into the digital end organ (as in the equine and bovine hoof) or form a separate digital pad (as in the canine claw, or the human finger tips or finger pulps). The same applies to the periople that - when comparing different types of digital end organs - may also stay 'separate' from the proper digital end organ and then rather form a perioplic crease. For the avian claw, an "unguical scale" adjacent to the "claw proper" (Moser, 1906; Lucas and Stettenheim, 1972) and a "subunguical scale" (Lucas and Stettenheim, 1972) adjacent to the digital pad ("terminal pad") have been described.

For clinical considerations, particularly the *innervation pattern and the angioarchitecture* of the avian claw are important.

The fundamental morphological study of Lucas and Stettenheim (1972) described abundant nerves within the claw proper, particularly in the solear dermis, besides lamellar corpuscles within the dermis of the dorsal plate of the claw. Intense sensitive innervation patterns with specialised nerve endings were also described for the ostrich foot pad (Palmieri *et al.*, 2003).

Regarding its vasculature, the avian claw is supplied by the arteries and veins of the claw bone, arising from the anterior tibial artery and from the venous tarsometatarsal rete, respectively (Vollmerhaus and Hegner, 1963). The basic study of Moser (1906) already described that the dermis of the distal scales and the claw is well vascularised and the vessels are arranged in horizontal plexuses.

Four distinct *vascular plexus*, listed in the Nomina histologica (1992), occur in the mature non-modified skin: The subcutaneous plexus, the profound and the superficial dermal plexus as well as the intrapapillary (or subepidermal) capillary plexus (Bravermann, 2000). In the highly derived pododerma of the mammalian digital end organ, the subepidermal vascular plexus is particularly well developed (Schummer, 1951; Dobler, 1969; Hirschberg *et al.*, 2001a). During the embryonic respectively foetal development of the human integument, both mesenchymal connective tissue layers, dermis and subcutis, are hardly distinguishable (Smith and Holbrook, 1982), and the vascular network is organized in one (early stage) or two planes parallel to the epidermis (Johnson and Holbrook, 1989). In later stages, a vascular plexus separates the dermis from the subcutis (Smith and Holbrook, 1982). Comparable findings have been reported for the foetal development of different mammalian end organs (equine hoof: Bragulla, 1996; canine claw: von Süsskind-Schwendi, 2005).

The prenatal development of the specific integumental structures of the avian claw therefore serves as an excellent model for studying the functional adaptation of the skin, as well as for a comparative morphological approach on the development and the phylogeny of digital end organs in general (Hirschberg *et al*, 2001b), while also rendering the prerequisite for evaluating the clinical implications of the claw.

In this study, the prenatal development of the avian claw was examined using light microscopic and scanning electron microscopic techniques on the feet of different developmental stages of cockatiel, duck, and chicken. The present preliminary report focuses on the development of the general structure of the avian digital end organ, i.e. the epidermis, the dermis, the subcutis, and the distal phalanx, and the formation of specific vascular plexuses as benchmarks for distinguishing presumptive dermal from presumptive subcutaneous layers in this process.

Materials and Methods

The claws of different developmental stages of 23 cockatiel foetuses (*Nymphicus hollandicus*; day 7 to day 20), of 9 domestic duck foetuses, hatchlings and neonates (*Anas platyrhynchos* forma *domestica*; day 24 to day 28, newly hatched) and of 5 adult chicken (*Gallus gallus* forma *domestica*) were used for the investigation of the development of the avian claw.

In the younger specimens, the whole foot was cut off and fixed in buffered formaldehyde solution (4 %; pH 7.2; 4 °C; 24 h). In the older stages, the claws were exarticulated in the distal interphalangeal joint and bisected in the sagittal plane. After fixation in buffered formaldehyde solution, the specimens were dehydrated in a graded series of alcohol and embedded in paraffin wax. For light microscopic examinations, cross and longitudinal serial sections (5 - 7 μm) of the digits (second, third and fourth digit) or claws were stained routinely with Hematoxylin and Eosin (HE), with Periodic acid-Schiff (PAS), or with trichrome (GRA) according to Romeis (1989).

In order to study the conformation of the dermal papillary body, the epidermal claw capsule was separated from the dermal surface by a special modified acetate-maceration technique (Bragulla, 1996). Whole non-macerated, as well as epidermal and dermal part of the macerated claw specimens were fixed in 2.5 % phosphate-buffered solution (pH 7.2) of formaldehyde, rinsed in phosphate buffered saline (pH 7.2) several times, dehydrated in a graded series of alcohol and dried using hexamethyldisilazane (Roth, Karlsruhe, Germany). The specimens were mounted onto aluminium stubs with Leit-C glue (Plano, Marburg, Germany) and sputter-coated with gold (3 min, 30 - 40 nm). Scanning electron microscopy (Nanolab 2000, Bausch & Lomb, Canada) was carried out at an accelerating voltage of 5 - 10 kV.

The chicken specimens employed in this study were collected from a local abattoir, the claw material of prenatal stages of cockatiel and duck was generously donated from other research projects (see acknowledgments).

Results

In all studied species, the digital end organ was characterised by a specifically modified skin, i.e., epidermis, dermis and subcutis, moulding a species-specifically formed cornified claw capsule produced by the keratinising and cornifying layers of the claw epidermis.

Structure of the adolescent and adult avian claw (figs. 1A and 1B)

The fully developed avian digital end organ was characterised by a cornified claw capsule covering the tip of the digit which was partly ensheathed by the adjacent distal scutate (dorsal) or reticulate (plantar) scales in the proximal part, i.e. the basis, of the claw, the latter thereby forming the claw crease. The inner aspects of both the dorsal and the plantar regions of the claw crease contributed to the cornified claw capsule and were thus termed the perioplic and the bulbar region of the claw, respectively. The hook-like shape of the claw was long and slender with a pointed tip and displayed a distinct dorsal curvature and a solear concavity. The length and the curvature of the avian claw varied species-specifically, but in all studied species a harder part covering the dorsolateral aspects of the claw, i.e. 'claw plate', and a softer 'solear' part of the claw capsule covering the plantar aspect was distinguishable (see fig. 1A).

The dermal surface as revealed by maceration-induced separation of the epidermis and the dermis (see fig. 1B) displayed a more pronounced pointiness of the claw, the formation of a dorsal ridge and of symmetrical lateral ridges in the dorso-lateral aspect as well as a slight transverse convexity and sagittal concavity of the plantar aspect of the avian claw.

The shape of the distal phalanx predominantly reflected the shape of the distal part of the claw proper, and displayed lateral vascular grooves (containing the proper digital arteries) that fused at the tip of the phalanx, thus forming an apical vascular channel situated deeply within the distal third of the distal phalanx - equivalent of the solear canal in the digital end organ of mammals - from which one or two osseous ridges (containing the apical branch/es of the terminal arch) originated towards the tip of the claw (compare to fig. 2 showing this feature in the maturing claw).

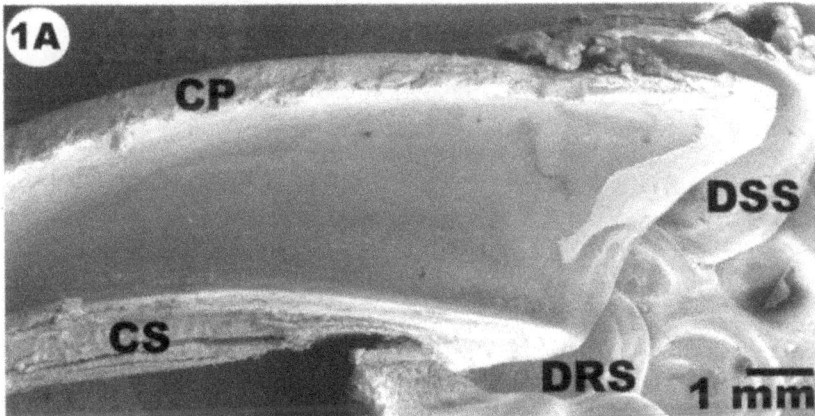

Fig. 1A. Structure of the adult avian claw (chicken): macerated specimen, inner surface of the epidermal cornified claw capsule, scanning electron microscopy.
CP = parietal part of the claw capsule (claw plate), CS = solear part of the claw capsule, DRS = distal reticulate scales, DSS = distal scutate scale

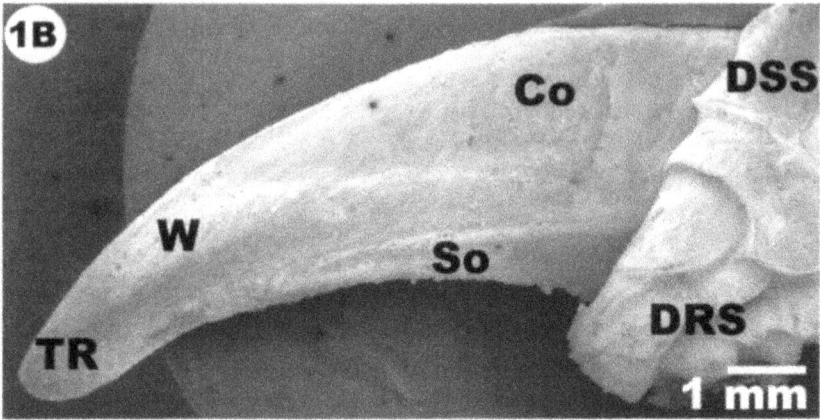

Fig. 1B. Structure of the adult avian claw (chicken): macerated specimen, dermal surface, scanning electron microscopy.
Co = coronary segment, DRS = distal reticulate scales, DSS = distal scutate scale, So = solear segment, TR = terminal region, W = wall proper

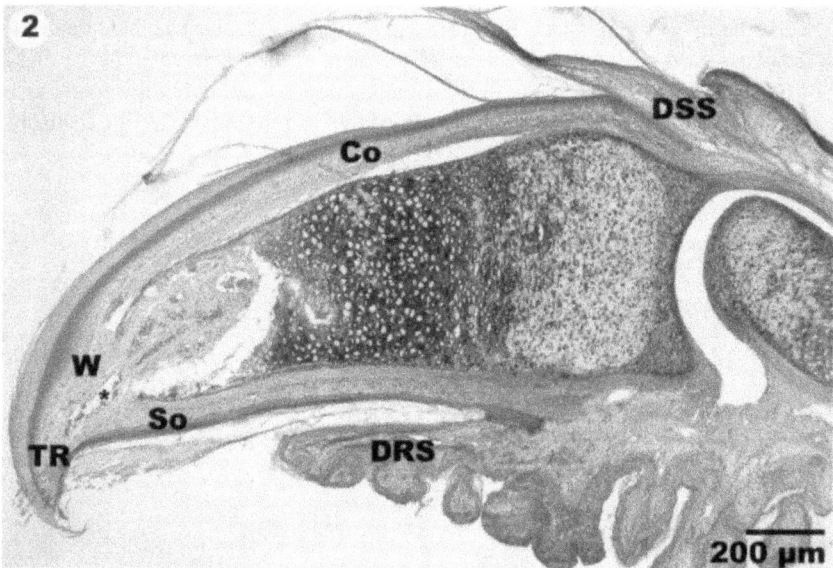

Fig. 2. Structure of the developing avian claw (cockatiel, day 20, third claw): light microscopy, trichrome stain (GRA), median section.

Co = coronary segment, DRS = distal reticulate scales, contributing to the cornified claw capsule (i.e. bulb), DSS = distal scutate scale, contributing to the cornified claw capsule (i.e. periople), So = solear segment, TR = terminal region, W = wall proper

At the end of the third trimester of gestation, near hatching, the species-specific shape of the claw is nearly fully developed. The distal phalanx is in part still cartilagineous, whereas only its apical part is fully ossified. Note the mass of flaky horn covering the tip and the solear part of the claw capsule, and the apical vascular branch (asterisk) originating from the profound arterial network of the claw encased in the ossified part of the distal phalanx.

Foetal development of the avian claw (figs. 2-6)

The development of the specialised and modified parts of the skin of the avian foot starts in the second trimester of gestation. The formation of the digital end organ was initiated by the occurrence of a slight thickening of the skin in the dorsal and plantar aspect of the softly rounded tip of the respective digit (fig. 3A). The anlage of the distal phalangeal bone was fully cartilaginous (primordial cartilaginous skeleton) and was capped by a distinct formation of (desmal) woven bone at the apical aspect of the rounded bone anlage. The next developmental stage was characterised by the lengthening of the digital end organ bud and a more pronounced ossification process at the tip of the distal phalangeal bone (fig. 3B). Plantar, a proliferation of loose connective tissue initiated the development of the bulb, whereas in the dorsal region the formation of a shallow crease indicated the development of the periople. At the end of the second trimester of gestation, the typical shape of the digital end organ with a pointed tip was already recognisable (see fig. 4A). The distal scutate and reticular scales ensheathing the proximal part, i.e. the basis, of the claw were developing. In this developmental stage, the tip of the distal phalangeal bone was completely osseous and an ossifying shell around the remaining part of the cartilaginous phalangeal anlage extended proximally. A thin claw capsule was produced consisting of a cornified plate and a sole part, respectively. The terminal and solear part of the cornified capsule displayed a mass of loose and flaky horn (figs. 2, 4B and 5).

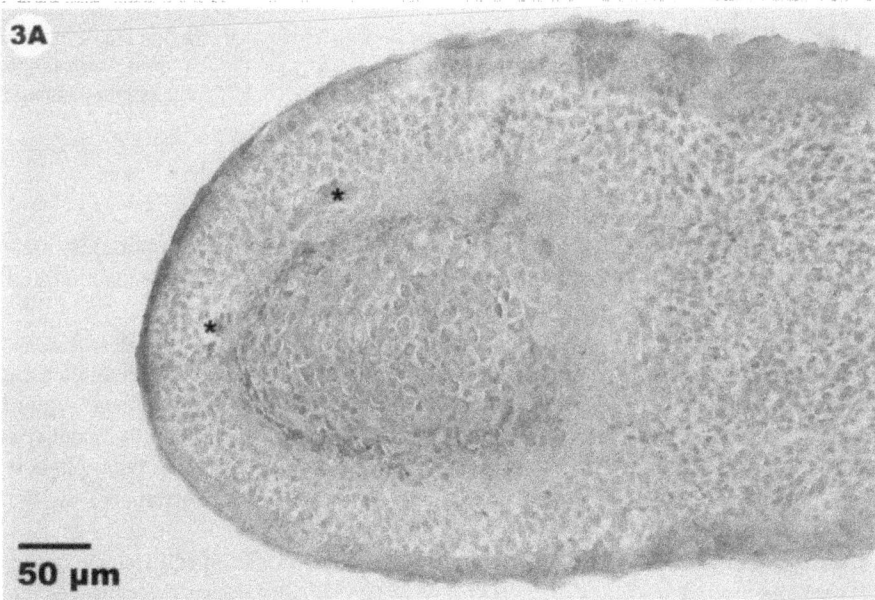

Fig. 3. Early development of the avian claw anlage in the second trimester of gestation (light microscopy):

A: Bud-like shape of the developing claw anlage, with a slight thickening of the epithelium; the distal phalanx is fully cartilaginous with only a slight margin of desmal woven bone at the tip of the claw anlage; note the capillary plexus supplying the claw, arranged in a single plane (asterisks). (cockatiel, day 8, PAS stain, longitudinal section of the third digit)

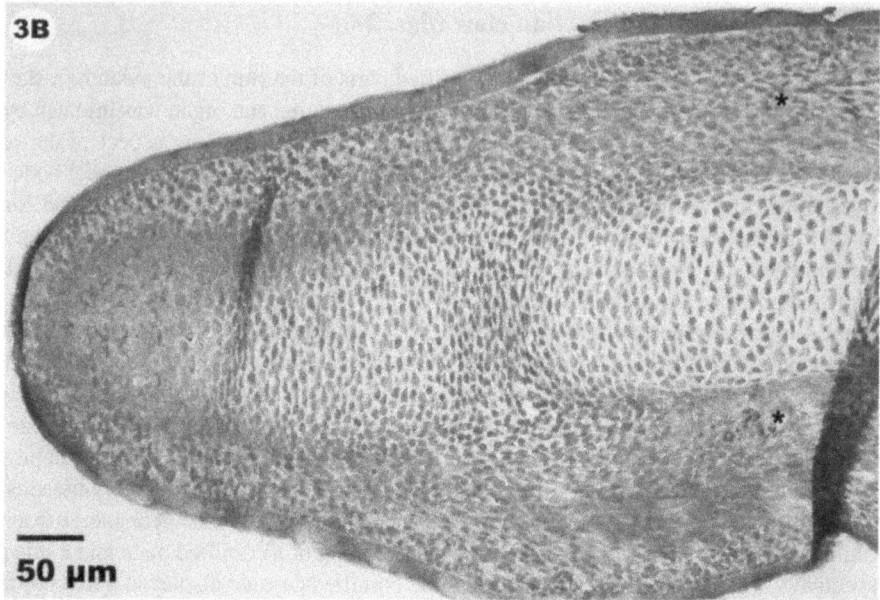

B: Elongation of the claw anlage; thickening of the dorsal (developing perioplic crease) and plantar (developing bulb) aspect; note the more prominent apical ossification of the tip of the distal phalanx and the still single-tier capillary plexus (asterisks). (cockatiel, day 10, HE stain, longitudinal section of the third digit)

At the beginning of the third trimester of gestation, the formation of a perioplic crease by enfolds of the distal scutate scales, and the formation of the bulb, i.e. the distal digital pad, by enfolds of the distal reticular scales, respectively, was definitive. The shape of the avian claw was now somewhat altered by the appearance of a slight ridge at the dorsal aspect and a more pronounced concavity at the plantar aspect of the claw. The cornified claw capsule was getting thicker with a mass of flaky horn covering the tip and the solear region. The inner aspects of the folds of both, distal scutate scales, i.e., periople, and distal reticular scales, i.e., bulb, contributed to the formation of the claw capsule. The distal phalanx was now widely ossified and developed a bone marrow containing (i.e., medullary) cavity in the distal third of the phalanx (see fig. 2).

At the end of the third trimester of gestation, the final, species-specific shape of the claw was developed (cockatiel: day 20, duck: day 24). A distinct dorsal ridge had developed, separating both lateral aspects of the claw which in turn had also developed low parallel ridges. The now nearly completely ossified distal phalanx displayed a corresponding shape with a dorsal ridge (fig. 5). The distal third of the dorsal part of the claw as well as the tip and the solear aspect of the claw were capped by thick masses of loose and flaky horn. Underneath this flaky horn mass, a thinner layer of compact horn was visible covering all regions of the claw (see figs. 2, 4B and 5).

In the early, less differentiated stages of claw development, only one vascular plexus was detectable in the mesenchyme of the distal scales and of the claw proper (fig. 3). With the development of the distinguishable claw anlage, two distinct super-imposed vascular plexus occurred within the mesenchymal layers, interconnected by ascending branches (fig. 4a). In the later stages of claw development, accompanied by the development of the formation of a distinct papillary body particularly in the scutate and reticular scales (fig. 6), a third peripheral capillary plexus was formed from which subepidermal capillary loops originated (figs. 4B, 4C). Major blood vessels from the deep vascular plexus supplying the digital end organ accompanied the proximal part of the anlage of the distal phalanx and formed a 'terminal arch' network at the apical end, thus dividing the ossified apical and the still cartilaginous part of the phalangeal anlage (figs. 4A, 4B). In the proximal coronary region of the claw proper, these deep major blood vessels flanked the dorsal ridge of the distal phalanx (fig. 5). These larger vascular networks were accompanied by small calibre nerves.

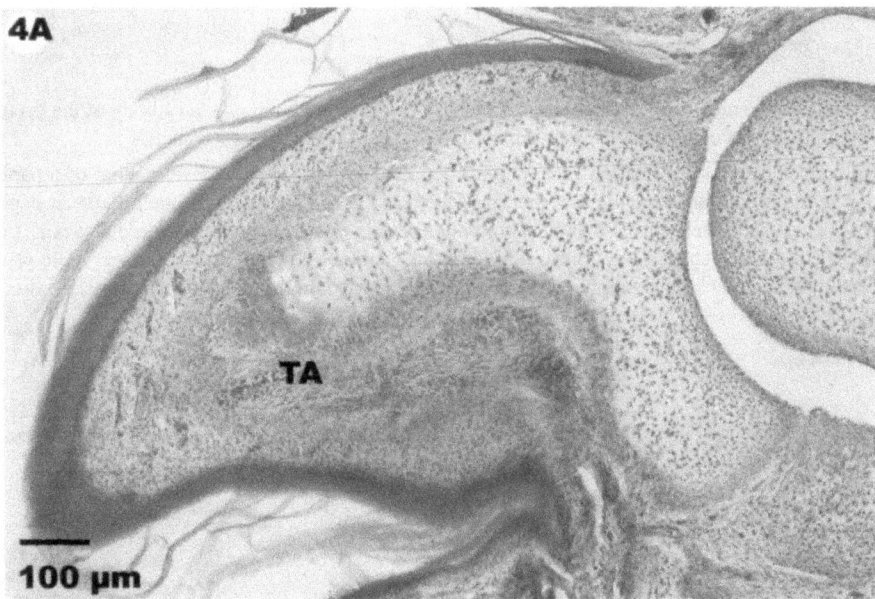

Fig. 4A. Formation of the 'terminal arch' (light microscopy, cockatiel, day 15, third digit, HE stain, slightly paramedian longitudinal section):

Distinct formation of two superimposed vascular plexus with ascending, interconnecting branches. At the tip of the claw anlage, the deeper plexus is encased within the ossified tip of the distal phalanx thus forming a 'terminal arch' (TA) while apical and solear ascending branches towards the wall and sole segment of the claw are detectable. Apart from the two-tier angioarchitecture, the connective tissue of the coronet, the wall proper and the terminal region of the claw proper appear undifferentiated.

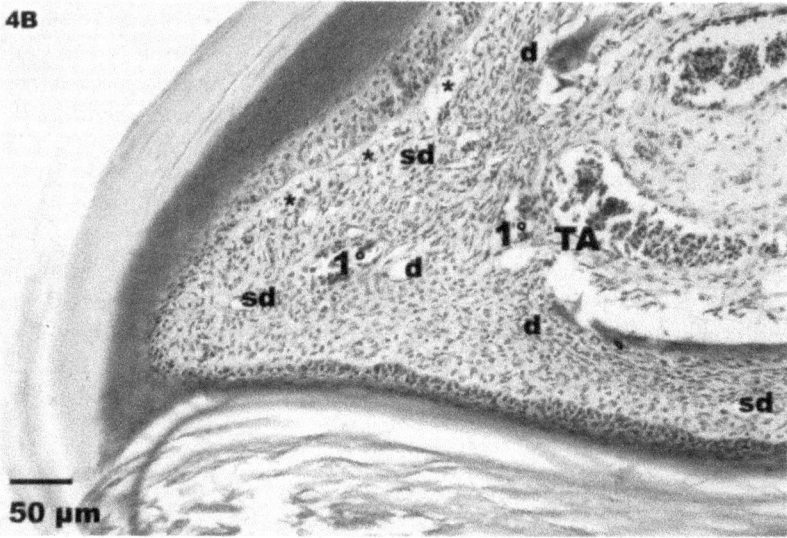

Fig. 4B. Development of the subepidermal plexus correlated with the papillary body at the tip of the claw (light microscopy, cockatiel, day 20, third digit, HE stain, longitudinal section):

Note the distinct wavy formation of the dermo-epidermal interface (i.e., development of a papillary body) in the distal dorsal area and at the tip of the claw, accompanied by a small diameter capillary plexus (asterisks) closely adjacent to the dermo-epidermal interface, i.e. the subepidermal plexus. Larger vessels ('terminal arch', TA) are encased within the ossified tip of the distal phalanx while primary ascending branches (1°) are supplying the deep dermal (d) and the superficial dermal (sd) plexus. Also note the thick masses of flaky horn encasing the tip and the sole of the claw proper.

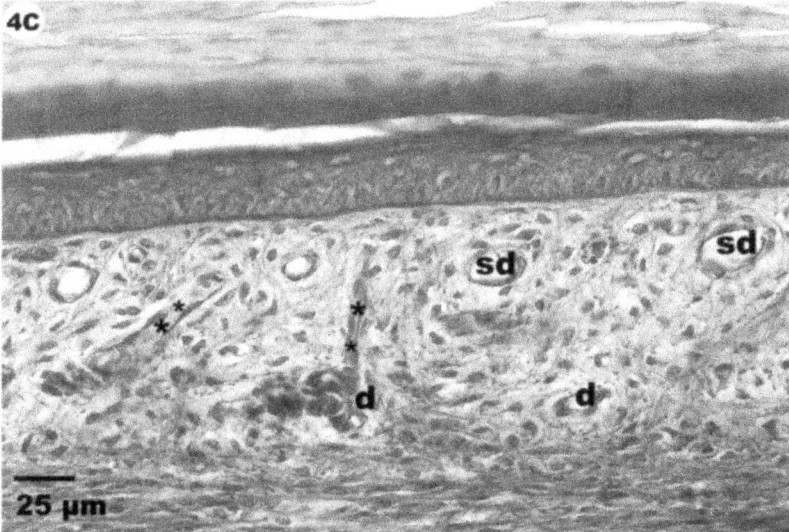

Fig. 4C. Dermal angioarchitecture of the proximal part of the coronet (duck, hatching day, second digit, GRA stain, longitudinal section):

Note the ascending capillary branches (asterisks) connecting the deep dermal (d) and the superficial dermal (sd) capillary plexus within the loose connective tissue layer in the proximal part of the coronet. While the dermo-epidermal interface remains still smooth, no subepidermal capillary loops are formed yet.

A 'typical' subcutaneous, i.e., loose connective tissue layer, was found in the distal scales, but not in the distal respectively apical regions (wall proper, terminal matrix, sole) of the claw proper (see fig. 2), whereas the dorsal proximal part of the claw (coronet) displayed a loose connective tissue cushion lateral to the dorsal ridge of the distal phalanx (fig. 5).

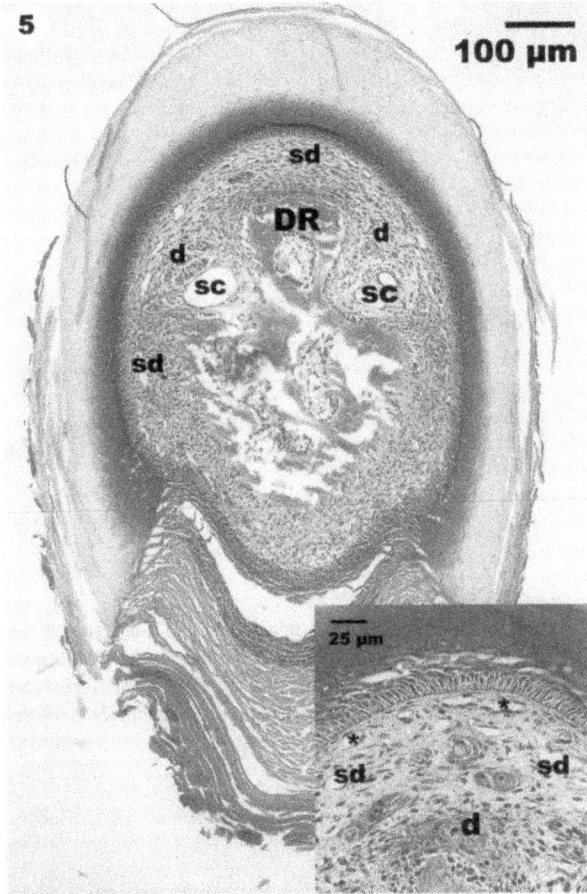

Fig. 5. Development of the coronary segment - vasculature and dorsal ridge (duck, day 28, second digit, HE stain, cross section; Insert: higher magnification of the connective tissue of the dorsal ridge, GRA stain):

Near hatching, the distal phalanx is widely ossified, including the basis of the dorsal ridge (DR), i.e. the distal part of the coronet. Note the wide vessels ('proper digital artery') flanking the dorsal ridge, representing the subcutaneous plexus (sc). The connective tissue core of the dorsal ridge is subdivided into a deeper dense layer and a superficial loose layer (insert). Three tiers of vascular cross sections are detectable, resembling the deep dermal (d), the superficial dermal (sd) and the subepidermal plexus (asterisks) - the latter presumably related to the commencing papillary body formation.

The mass of flaky horn covering the definite horn of the claw plate and - more prominently - the claw sole forms the deciduous claw capsule.

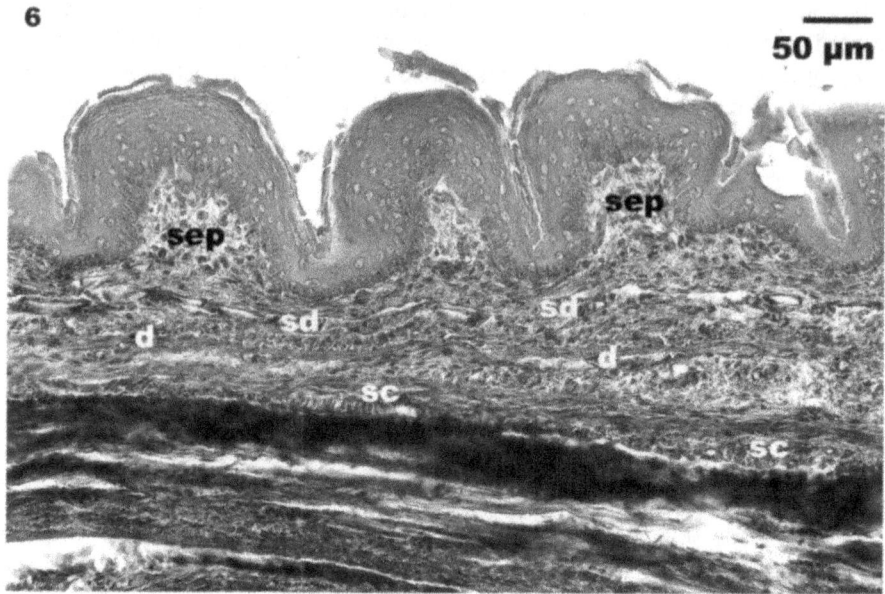

Fig. 6. Angioarchitecture of the reticulate scales (duck, day 28, fourth digit, GRA stain, longitudinal section):

The connective tissue of the reticulate scales is comprised of a deep layer associated with adjacent tendons or bones, respectively (i.e., subcutis), a denser, fibre-rich layer (reticular layer of the dermis) and a superficial cell-rich loose layer (superficial layer of the dermis). The dermo-epidermal interface is invaginated, forming a distinct papillary body. Associated with this function-related differentiation of the connective tissue, the angioarchitecture is developed accordingly, thus displaying a subcutaneous (sc), a deep dermal (d), a superficial dermal (sc) and a subepidermal (sep) plexus.

Discussion

This preliminary report is part of a detailed study on the structure and the development of the avian claw, respectively of a comparative study on the morphology and the phylogeny of the claw in general.

In the avian claw, the distal scales were contributing to the claw capsule formation, and - comparable to the situation in the mammalian digital end organ - were therefore termed perioplic respectively bulbar region of the claw in a broader sense, according to the allocation of segments described by Zietzschmann (1918).

During its prenatal development, the epidermis of the claw produced different generations of cornifying cells, thus forming a deciduous claw capsule ensheathing the permanent horn capsule ("epitrichium" of older studies; Moser, 1906). Like in mammals (Bragulla, 1998), the deciduous claw capsule is formed in the foetal period of the claw development in birds, whereas permanent claw horn is produced peri- and postnatally. The deciduous claw capsule covers the sharp tip and rim of the claw and thus protects the delicate foetal membranes during incubation.

The results of the present study confirmed older studies (Moser, 1906; Lucas and Stettenheim, 1972) concerning the abundance of nerves within the claw proper. Regarding

the common practice of declawing with microwave energy in commercial poultry production, it is therefore questionable whether this intervention is *de facto* really as painless as
commonly assumed; and a more detailed examination of the quality of these claw nerves is
thus urgently needed.

The development of the pododermal vasculature of the cockatiel and duck was comparable to that described in humans (Smith and Holbrook, 1982; Johnson and Holbrook, 1989).
A fully differentiated angioarchitecture including a subepidermal (or intrapapillary) plexus
was only developed in those regions of the claw proper that displayed a distinct papillary
body. Like in the mammalian end organ (Bragulla and Hirschberg, 2003), formation of the
papillary body seemed related to the course of the superficial dermal vasculature, with the
capillaries taking on a 'guide rail' function for the invagination of the dermo-epidermal
interface.

As reported by earlier studies (Moser, 1906; Lucas and Stettenheim, 1972), a 'typical'
subcutaneous layer was found in the distal scutate and reticulate scales and in the proximal
coronary region of the claw, but not in the distal regions of the claw proper, i.e., in the wall
and sole segments of the claw.

A separate desmal ossification of the tip of the distal phalanx was detected similar to
what has been described in the developing digital end organ of the horse (Carlens, 1927).
The development of specific vascular plexus in the connective tissue compounds of the claw
anlage separated the prospective dermal and subcutaneous layers even in the early, otherwise less differentiated stages of claw development (e. g., refer to fig. 4A). The established
occurrence and location of the subcutaneous and the profound dermal vascular plexus of-
fered a reliable benchmark to distinguish between the two connective tissue layers of the
claw integument. The 'proper digital artery' network (fig. 5) and the 'terminal arch' network
of the claw (fig. 4) resemble the equivalent of the subcutaneous plexus. Thus, ossification of
subcutaneous tissue in the distal respectively in the apical area, i.e. the wall proper and its
terminal region as well as the sole, of the developing avian claw was detected. At least in the
duck, the base of the dorsal ridge was also ossified, thus equally resembling subcutaneous
desmal bone formation.

Apparently, even the hitherto 'neglected' subcutaneous layer of the digital end organ is
highly modified (Hirschberg *et al.*, 2005). Whereas in the scales, the subcutis fulfils a 'typical' subcutaneous role in the form of a loose connective tissue layer, within the claw proper
the subcutis is in part ossified and fuses with the distal phalanx in order to allow a force-
transmission to the remaining structural parts of the claw. This specific modification of the
subcutis of the claw proper is presumably related to the specific functions of the claw integumental layers, i.e., up-take and transformation of weight- or strain-induced forces occurring during weight-bearing and/or in the process of climbing, digging or grasping.

The peculiarity of the development and of the ossification process of the distal phalanx
has been described for the mammalian digital end organ (reviewed by Hamrick, 2001),
while so far it has not been associated with formation of (sub)cutaneous bone. The homeobox
genes Msx1 and Msx2 are held responsible for provision of the inductive signal for terminal
phalanx ossification in the mammalian distal limb appendages, and the Bone Morphogenic
Proteins (BMPs) - prominent inducers of bone formation - regulate Msx expression. Current
evidence also suggests that Msx1 initiates membranous ossification in craniofacial elements
(review: Hamrick, 2001).

These insights into principles of modified skin development are essential for evaluating
the pathogenesis of major skin problems in poultry husbandry such as pododermatitis (Breuer

et al., 2005), and provide more data for a comparative morphological approach on evolution and phylogeny of the claw in general.

Acknowledgements

The authors wish to thank our technical assistants Karin Briest-Forch and Ilona Küster-Krehahn for their excellent support. We also acknowledge the general support of this study by our former doctorate student Dr. Marietta Freiin von Süsskind-Schwendi (Berlin, Germany), as well as by PD Dr. Sven Reese (Ludwig-Maximilians-University, Munich, Germany; cockatiel specimens) and Prof. Dr. Karl-Dietrich Weyrauch (Freie Universität, Berlin, Germany; duck specimens), the latter particular for provision of the avian material employed in this study.

References

ANONYMOUS (2006) Hoatzin. In: Wikipedia, Online Encyclopedia; http://en.wikipedia.org/wiki/Hoatzin; 02.05.2006

BOAS, J. E. V. (1931) Krallen (inkl. Nägel, Hufe, Klauen). In: *Bolk, L., E. Göppert, E. Kallius, and W. Lubosch* (eds.): Handbuch der vergleichenden Anatomie der Wirbeltiere. Berlin, Urban and Schwarzenberg, 521-544

BRAGULLA, H. (1996) Zur fetalen Entwicklung des Pferdehufes. (On the fetal development of the equine hoof.) Berlin, Faculty of Veterinary Medicine, Freie Universität, Habilitation thesis.

BRAGULLA, H. (1998) Zur pränatalen Entwicklung der Hufkapsel. (Prenatal development of the hoof capsule.) *Wien Tierärztl Mschr* **85**, 233-244

BRAGULLA, H. (2003) Fetal development of the segment-specific papillary body in the equine hoof. *J Morph* **258**, 207-224

BRAGULLA, H., AND R. M. HIRSCHBERG (2003) Horse hooves and bird feathers: Two model systems for studying the structure and development of highly adapted integumentary accessory organs - The role of the dermo-epidermal interface for the microarchitecture of complex epidermal structures. *J Exp Zool,* **298B**, 140-151

BRAVERMANN, I. M. (2000) The cutaneous microcirculation. *J Investig Dermatol Symp Proc* **5**, 3-9

BREUER, P., S. BUDA, AND K.-D. BUDRAS (2005) Ultrastructural study of the pre- and postnatal development of the foot pads of turkeys – regarding clinical implications for the occurrence of foot pad lesions. Proc. 2nd Combined Workshop Fundamental Physiology of the European Working Group of Physiology and Perinatal Development in Poultry, Berlin, 23rd-25th September 2005. Humboldt Universität zu Berlin, 14

CARLENS, O. (1927) Beitrag zur Kenntnis der embryonalen Entwicklung des Extremitätenskelettes beim Pferd und Rind: Teil I und II. (Contribution to the knowledge on the embryonal development of the skeleton of the extremeties in horses and cattle: Part I and II). *Morph Jb* **58**, 153-196 a. 367-412

COMPTON, M. M., H. P. VAN KREY, P. L. RUSZLER, AND F. C. GWAZDAUSKAS (1981) The effects of claw removal on growth rate, gonadal steroids, and stress response in cage reared pullets. *Poult Sci* **60**, 2120-2126

DOBLER, C. (1969) Papillarkörper und Kapillaren der Hundekralle, Schweine- und

Ziegenklaue. (Papillary body and capillaries of the canine claw and the porcine and the caprine hoof.) *Gegenbaurs Morphol Jb* **113**, 382-428

FRANKENHUIS, M. T, M. H. VERTOMMEN, AND H. HEMMINGA (1991) Influence of claw clipping, stocking density and feeding space on the incidence of scabby hips in broilers. *Br Poult Sci* **32**, 227-230

HAMRICK, M. W. (2001) Development and evolution of the mammalian limb: adaptive diversification of nails, hooves, and claws. *Evol Develop* **3**, 355-363

HARGIS, B. M., R. W. MOORE, AND A. R. SAMS (1989) Toe scratches cause scabby hip syndrome lesions. *Poult Sci* **68**, 1148-1149

HIRSCHBERG, R. M., C. K. W. MÜLLING, AND K.-D. BUDRAS (2001a) Pododermal angioarchitecture of the bovine claw in relation to form and function of the papillary body. A scanning electron microscopic study. *Microsc Res Tech* **54**, 375-385

HIRSCHBERG, R. M., H. BRAGULLA, AND M. VON SÜSSKIND-SCHWENDI (2001b) The development of the avian claw. A comparative morphological approach. *J Morph* **248**, 241

HIRSCHBERG, R. M., M. VON SÜSSKIND-SCHWENDI, H. BRAGULLA, AND K.-D. BUDRAS (2005) Form follows function - Neue Erkenntnisse über das Unterhautgewebe des Zehenendorgans. (Form follows function - new insights into the subcutaneous tissue of the digital end organ.) Proceeding, 26. Kongress der DVG, DVG Service GmbH, Gießen, 97

HOMBERGER, D.G. (2000) The case of the cockatoo bill, horse hoof, rhinoceros horn, whale baleen, and turkey beard: The integument as a model system to explore the concepts of homology and non-homology. In: H. M. Dutta & J. S. Datta Munshi, eds., Oxford & IBH Publishing Co.: *Vertebrate functional morphology: Horizon of research in the 21st century*. New Delhi, New Hampshire, Enfield and Science Publishers Inc., 317-343

HONAKER, C. F., AND P. L. RUSZLER (2004) The effect of claw and beak reduction on growth parameters and fearfulness of two Leghorn strains. *Poult Sci* **83**, 873-881

JOHNSON, C. L., AND K. A. HOLBROOK (1989) Development of human embryonic and fetal dermal vasculature. *J Invest Dermatol 93* (Suppl 2), 10S-17S

LUCAS, A. M., AND P. R. STETTENHEIM (1972) Digital Claw. In: Lucas, A. M., and P. R. Stettenheim: Avian Anatomy. Part II. Agricultural Handbook 362, U.S. government. Washington D. C., Printing Office, 606-609

MOSER, E. (1906) Die Haut des Vogels. In: Ellenberger, W. (ed), *Handbuch der vergleichenden Anatomie der Haustiere*, Bd. 1, 192-232

NOMINA ANATOMICA VETERINARIA (2005) International Committee on Veterinary Gross Anatomical Nomenclature, World Association of Veterinay Anatomists. 5th edition. Editorial committee, Hannover, Columbia, Gent, Sapporo

NOMINA HISTOLOGICA (1992) Revised 2nd edition. Department of Veterinary Anatomy, Cornell University, Ithaca, New York, USA

PALMIERI, G., M. SANNA, L. MENELLI, M. BOTTI, F. GAZZA, A. DI SUMMA, N. SANTAMARIA, L. PASSANTINO, M. MAXIA, AND F. ACONE (2003) On the sensitive innervation of the ostrich's foot pad. *Ital J Anat Embryol* **108**, 25-37

ROMEIS, B. (1989) Mikroskopische Technik,17th ed. Munich, Urban and Schwarzenberg

SANGIORGI, S., A. MANELLI, T. CONGIU, A. BINI, G. PILATO, M. REGUZZONI AND M. RASPANTI (2004) Microvascularisation of the human digits studied by corrosion casting. *J Anatomy* **204**, 123-131

SCHUMMER, A. (1951) Blutgefäße und Zirkulationsverhältnisse im Zehenendorgan des Pferdes. (Blood vessels and circulation in the digital end organ of the horse.) *Morphol Jb* **91**, 568-649

SMITH L. T., AND K. A. HOLBROOK (1982) Development of dermal connective tissue in human embryonic and fetal skin. *Scan Electr Microsc* **4**, 1745-51

SPEARMAN, R. I. C. (1985) Integument. In: A. S. King and J. MacLelland, eds.: *Form and function in birds. Vol. 3*, Chap. 1. London Academic Press, 1-56

VANSKIKE, K. P., AND A. W. ADAMS (1983) Effects of declawing and cage shape on productivity, feathering, and fearfulness of egg-type chicken. *Poult Sci* **62**, 708-711

VOLLMERHAUS, B., AND D. HEGNER (1963) Korrosionsanatomische Untersuchungen am Blutgefäßsystem des Hühnerfußes. (Corrosion cast anatomical study of the vascular system in the foot of the fowl.) *Morph Jb* **105**, 139-184

VON SÜSSKIND-SCHWENDI, M. (2005) Die prae- und perinatale Entwicklung der Hundekralle. (The pre- and perinatal development of the canine claw.) Berlin, Faculty of Veterinary Medicine, Freie Universität, dissertation thesis. Mensch & Buch, Berlin

ZIETZSCHMANN, O. (1918) Das Zehenendorgan der rezenten Säugetiere: Kralle, Nagel, Huf. (The digital end organ of extant mammals: claw, nail, hoof.) *Schweizer Archiv für Tierheilkunde* **60**, 241-271

Investigations of the pre- and postnatal development of the foot pad skin of turkey poults

PIA BREUER, SILKE BUDA* and K.-D. BUDRAS

Institute of Veterinary Anatomy, Faculty of Veterinary Medicine, Freie Universität Berlin, Koserstraße 20, D-14195 Berlin, Germany

Although foot pad lesions are a common finding in commercial turkeys at slaughter, little is known about the onset of the disease. To elucidate possible causes of this alteration in growing meat type turkeys of a heavy breed (BUT Big 6), the development and fine structure of the foot pad skin region of foetuses and very young poults of turkey breeder hens at the beginning and at the end of the laying period were examined. No alterations of the foot pad skin were found in the turkey foetuses. However, at the age of four days, hyperaemia - the first symptom of a pathological alteration - was detected at the metatarsal foot pads. The progeny of turkey hens at the end of the reproduction period displayed a greater number and more severe cases of the skin irritation than the offspring of hens at the beginning of the laying period. Until the age of 7 days, the body weight of the older turkey hens' progeny was significantly higher than that of the younger layers, suggesting that in the first week of life a high body weight gain seems to have an exceptionally negative effect on maturing and adapting processes of the plantar skin. With little differences the pre- and postnatal development and fine structure of the metatarsal reticulate scale epidermis of turkeys is comparable to that of chickens.

Keywords: Turkey; reticulate scales; periderm; foot pad lesions

Introduction

In fattening turkeys severe cases of foot pad lesions are found frequently at slaughter (Platt, 2004). These lesions are the final stage of an inflammation starting at the plantar aspect of the foot (metatarsal and digital pads). Many factors like poor management, litter-quality (Clark *et al.*, 2002; Harms *et al.*, 1977; Jensen, 1985), nutritional imbalances, diet composition (Austic and Scott, 1984; Balnave, 1970; Harms and Simpson, 1975; Roland and Edwards, 1971) and genetic predisposition of turkey lines (Hafez *et al.*, 2004) have been discussed to cause foot pad dermatitis (pododermatitis). These skin damages are likely to function as initial events of a spreading inflammation that might lead to further serious alterations like leg weakness or septicaemia, affecting the whole body. Considering this, foot pad dermatitis is not only an economic problem that causes losses in the rearing period and impairment of carcass quality (Ekstrand and Algers, 1997). Because of the pain from which the animals are

*Corresponding author: Dr. Silke Buda
E-mail: buda.silke@vetmed.fu-berlin.de

likely suffering (Ekstrand *et al.*, 1998), pododermatitis is also an animal welfare issue. Although the relevance of these alterations is evident, little is known about the beginning, not to mention the early signs of the disease and its predisposing factors. Therefore, the pre- and postnatal development of the reticulate scales that cover the plantar surface of the turkey's foot (and of the modified skin of the avian foot in general – see also Hirschberg *et al.*, 2005) is of particular interest.

Material and methods

For this purpose 70 birds from the offspring of BUT Big 6 turkey breeders (heavy meat type turkeys) kept under field conditions were examined macro- and microscopically at the beginning (production week (PW 4)) and the end (PW 20) of the reproduction period, respectively. The investigation was carried out on five randomly taken birds from different pre- (incubation day (ID) 20, 23, and 26) and postnatal developmental stages (newly hatched (ID 28), day 7, 14, and 21 of life). From all stages the body-weight was recorded, while from ID 20, 23 and 26 the egg-weight was documented, additionally. After culling, tissue specimens of the metatarsal foot pads were collected from every developmental stage. All samples were processed with routine histological and scanning electron microscopic techniques.

The postnatal investigations (day 7, 14, and 21 of life) took place on male poults that were reared under experimental conditions. The housing occurred under climate-, air- and light-controlled conditions on non-impregnated softwood shavings and long-straw. During the investigation period, the turkey poults had free access to the round drinkers and feeders, while they received a commercial turkey diet in two phases. Daily the foot pad condition was monitored.

MACROSCOPIC SCORING OF TURKEY FOOT HEALTH STATUS

The health status of the metatarsal foot pads of the turkey poults was assigned in four grades: 0 = sound; 1 = redness; 2 = brown discoloration; 3 = brown discoloration and hyperkeratosis (increased formation of corneal cells). This gradation refers to the lesion score for adult turkeys used at slaughter described by Ekstrand and Algers (1997). A modification was necessary to consider the particular conditions of alterations in very young birds.

Results

WEIGHT OF TURKEY EGGS, FOETUSES AND POULTS

With the exception of day 23 of incubation, eggs from PW 20 were - with a mean weight of 84.3 g - statistically significantly heavier than those of PW 4, by an average of 4.6g.

Until the age of 7 days, the body weight of the older turkey hens' progeny was statistically higher than that of the younger layers. Until hatch the offspring of PW 20 were on average 5.4 g heavier than those from PW 4. After hatch on day 7 of life the older turkey hens' progeny is on an average of 36.1 g heavier than that of PW 4. On day 14 and 21 of life no statistically noticeable difference in mean body-weight could be found (Table 1).

Table 1. Mean body-weight of turkey foetuses and poults in grammes

Day of incubation/life	Production week 4	Production week 20
20	47.0	52.5
23	60.4	63.0
26	62.1	73.8
28	56.9	58.7
7	155.4	191.5
14	454.6	422.7
21	805.5	787.1

n = 5

MACROSCOPIC EXAMINATION AND FOOT PAD HEALTH STATUS

Foetuses and newly hatched turkeys (ID 28) exhibit the expected skin morphology with reticulate scale formation on the metatarsal and digital foot pads. Poults originating from the breeders late in production (PW 20) revealed first symptoms of a foot pad dermatitis at the age of four days, starting with a redness of the metatarsal pads (Fig. 1). From the first to the third week of life, 27 % of the progeny from the younger breeders (production week 4) and 80 % of the progeny from the older breeders displayed foot pad alterations of grade 1-3. The offspring of breeders from production week 4 revealed pathologic metatarsal pad affections mainly of grade 1 and 2, displaying a tendency to grade 2 with increasing age, while the progeny of breeders in production week 20 showed mainly alteration signs of grade 2 and 3, tending towards grade 3 (Fig. 2).

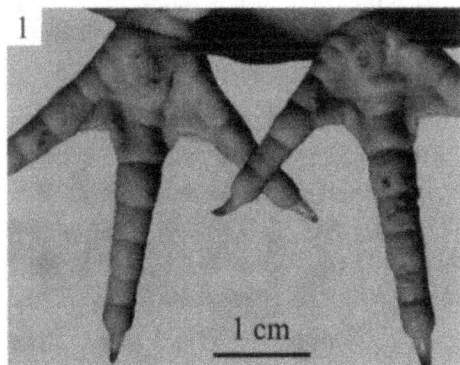

Figure 1. Hyperaemia of the metatarsal foot pads of a 4 days old turkey poult

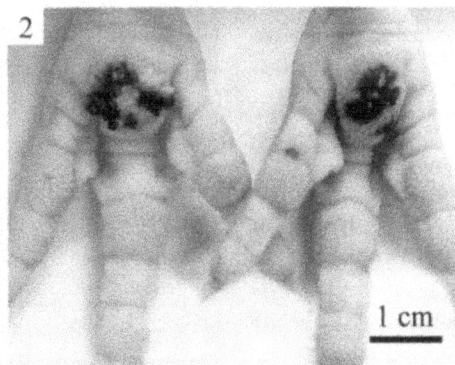

Figure 2. Dyscoloration and hyperceratosis of the metatarsal pad of a 3 weeks old turkey

STRUCTURE OF METATARSAL FOOT PADS OF FOETUSES AND YOUNG TURKEY POULTS

In order to investigate the development and physiological structure of foot pads of foetuses and young turkey poults, all samples were taken from unaltered areas of the foot pad skin.

HISTOLOGICAL AND SCANNING ELECTRON MICROSCOPICAL OBSERVATIONS

Day 20 of incubation

The reticulate scales - regular and cuboidal structures - were localised on the foetal metatarsal pads. In the dermal layer, many capillaries were situated right beneath the dermo-epidermal interface. The epidermal stratum basale was formed by a single layer of columnar cells being anchored on the slightly undulated basement membrane that - in turn - represented the transition from dermis to epidermis. The stratum basale was followed by the intermediate layer. As they moved towards the skin surface, the intermediate cells were distinctly flattened and the wide intercellular clefts became narrow. The border between foetal epidermis and the amniotic fluid was formed by peridermal cells. Their outer surface was covered by microfold-like protuberances that looked similar to those of the human fingertip.

Day 23 of incubation

Underneath the periderm a narrow foetal stratum corneum had developed. Consequently, the peridermal cells lost their nutritional and protective function and were sloughed off.

Day 26 and 28 of incubation

From day 26 to hatch at day 28 of incubation, the stratum corneum thickened intensely. On the outer surface of the foot pad epidermis, the corneal cells were sloughed off singly or in small conglomerates. In contrast to the cells of the stratum corneum, the cells of the living part of the epidermis exhibited many adhesive cell contacts (desmosomes). By passing through cell differentiation the number of these junctions decreased. In young cornified cells the desmosomes to the neighbour-cells were limited to the lateral cell margins, the basal and apical aspects of the cells were almost plane. Mature cornified cells had almost lost the intercellular contact, thus facilitating their slough-off.

Day 7 of life

In contrast to the foetal reticulate scales, at this developmental stage the scales appeared narrowly spaced, thus giving the impression of a tight metatarsal surface. Not only were intact cornified cells sloughed off, they also declined on the scale surface. Because of this process of disintegration, the lipid content of the mature corneal cells was liberated onto the surface of the metatarsal pad.

Day 14 and 21 of life

The structure of the skin corresponded to that described for day 7 of life.

Discussion

In its main features the development and structure of the reticulate scale epidermis of the BUT Big 6 turkeys examined in this study is comparable to that of chicken and adult turkeys (Wäse, 1999; Platt, 2004). However, some variations such as the disintegration and lipid-liberation of the mature cornified cells on the metatarsal surface were detectable. The peridermal protuberances detected in the present investigation correlate with peridermal microvilli described for human embryos (Holbrook and Odland, 1975) and have not been described for turkey foetuses before. Via enlargement of the skin surface, the microfolds are thought to be responsible for resorption of nutrients from the amniotic fluid (Holbrook and Odland, 1975). The periderm itself seemed to have a protective function as long as a foetal stratum corneum has not yet been formed.

No morphological differences in the metatarsal skin structure between the progeny of production week 4 and 20 were found. Nevertheless the offspring of turkey hens late in reproduction (PW 20) generally had a poorer health condition regarding the metatarsal pads compared to the progeny of younger turkey hens (PW 4). In the first week of life, a high body weight gain seemed to have an exceptionally negative effect on the foot pad condition of the poults. The adaptation and maturing processes of the plantar skin are probably disturbed by an increased mechanical load, leading to the poor foot pad condition of the poults originating from the older hens.

Acknowledgements

This study is part of the dissertation thesis of Pia Breuer (Faculty of Veterinary Medicine, Freie Universität Berlin, 2005) and was financially supported by the Senate of Berlin (NaFöG-stipend). The authors also wish to thank Roche Vitamins Ltd. and Moorgut Kartzfehn for their support.

References

AUSTIC, R.E. and SCOTT, M.L. (1984) Biotin. *Diseases of Poultry*, 8th ed, HOFSTAD, M.S., BARNES H.J., CALNEK B.W., REID W.M. and YODER H.W., Jr (eds.), American Association of Avian Pathologists, Iowa State University Press, Ames, Iowa, USA, 54-55.

BALNAVE, D. (1970) Essential fatty acids in poultry nutrition. *World's Poultry Science Journal* 6: 42-460.

CLARK, S., HANSEN, G., McLEAN, P., BOND, P. JR., WAKEMAN, W., MEADOWS, R. and BUDA, S. (2002) Pododermatitis in turkeys. *Avian Diseases* **46(4)**: 1038-1044.

EKSTRAND, C. and ALGERS, B. (1997) Rearing conditions and foot-pad dermatitis in swedish turkey poults. *Acta Veterinaria Scandinavia* **38**: 167-174.

EKSTRAND, C., CARPENTER, T.E., ANDERSSON, I. and ALGERS, B. (1998) Prevalence and control of foot-pad dermatitis in broilers in Sweden. *British Poultry Science* **39**: 318-324.

HAFEZ, M.H., WÄSE, K., HAASE, S., HOFFMANN, T., SIMON, O. and BERGMANN, V. (2004) Leg disorders in various lines of commercial turkeys with especial attention to pododermatitis. *Proceedings of the 5th international symposium on turkey diseases*, Berlin (ed. HAFEZ, H.M.), DVG-Service GmbH Verlag, Giessen, Germany, 11-18.

HARMS, R.H. and SIMPSON, C.F. (1975) Biotin deficiency as a possible cause of swelling and ulceration of foot pads. *Poultry Science* **54**: 1711-1713.

HARMS, R.H., DAMRON, B.L. and SIMPSON, C.F. (1977) Effect of wet litter and supplemental and/or whey on the production of foot pad dermatitis in broilers. *Poultry*

Science **56:** 291-296.

HIRSCHBERG, R.M., BRAGULLA, H. and BUDRAS K.-D. (2005) The prenatal development of the avian claw – a model for the formation of functional-related modified skin. In: *New Insights into Fundamental Physiology and Peri-natal Adaptation of Domestic Fowl.* Nottingham University Press, Nottingham, pp151-166.

HOLBROOK, H.K. and ODLAND, G.F. (1975) The fine structure of developing human epidermis: Light, scanning and transmission electron microscopy of the periderm. *Journal of Investigative Dermatology* **65:** 16-38.

JENSEN, L.S. (1985) Effect of cage floor and diet on the incidence of pododermatitis and on health in broiler breeder males. *Abstract Poultry Science* **64** *(Supplement 1):* 122.

PLATT, S.L. (2004) Die reticulate scales an den Fußballen schwerer Mastputen und deren Beeinflussung durch unterschiedliche Biotindosierungen unter Feldbedingungen. (Investigation of the structure of reticulate scales from the foot pads of heavy-breed turkeys and the influence of the different biotin levels under field conditions.) Freie Universität Berlin, Faculty of Veterinary Medicine, Dissertation thesis.

ROLAND, D.A. and EDWARDS, H.M. (1971) Effect of essential fatty acid deficiency and type of dietary fat supplementation on biotin-deficient chicks. *The Journal of Nutrition* **101:** 811-818.

WÄSE, K. (1999) Studie über die gesunde Haut von Masthühnern und ihre Veränderungen bei einem experimentell erzeugten Biotinmangel. (Study on the healthy broiler chicken skin and its alterations under experimental biotin - deficiency -.) Freie Universität Berlin, Faculty of Veterinary Medicine, Dissertation thesis.

Increased testosterone content in egg differentially influences behaviour of Japanese quail during ontogeny

M. OKULIAROVÁ[1], L. KOSTAL[2], M. ZEMAN[1,2]

[1]Department of Animal Physiology and Ethology, Faculty of Natural Sciences, Comenius University, Mlynska dolina B-2, Bratislava, Slovak Republic, [2]Institute of Animal Biochemistry and Genetics, Slovak Academy of Sciences, Ivanka pri Dunaji, Slovak Republic

In several avian species it was shown that maternal androgens in yolk may represent one of the possible ways by which mothers can influence early development and behaviour of their offspring. Consistent with this view we examined behaviour of precocial Japanese quail during the ontogenetic period from hatching to 37 days of age. Before incubation we manipulated yolk androgen levels by injection of physiological dose of testosterone (25 ng) dissolved in an olive oil. Eggs of the control group were injected with 20 µl of olive oil. Subsequent observations included males only and we recorded their behavioural responses in four open-field tests that were performed at 2, 9, 23 and 37 days of age. No behavioural differences were found between groups in two day old quail chicks. However, on day 9 after hatching birds from testosterone-treated eggs displayed higher locomotor activity, vocalization and pecking behaviour that were accompanied by both shorter latency to vocalize and shorter latency to explore as compared to control birds. This pattern was evident in latency to explore and latency to vocalize at 23 days of age and in locomotor activity and latency to explore at 37 days of age. Our results suggest that increased levels of testosterone experienced during embryonic development may contribute to individual behavioural differences in postnatal life.

Key words: androgen, bird, maternal hormones, development

Introduction

Transgenerational transfer of hormones from mother to eggs and subsequently to embryos may represent important epigenetic adaptation to prevailing environmental conditions. Steroid hormones are considered important in this process because of their genomic action. Avian eggs contain maternal hormones that can influence development and behaviour of offspring (Schwabl, 1996). This phenomenon has been studied preferentially in free-living altricial birds that are completely dependent on parental care. The results suggest that maternal androgens influence growth, behaviour, immune functions and offspring survival (Groothuis *et al.*, 2005).

There are limited data on this subject in precocial birds, including poultry. Data published till now indicate that experimentally enhanced testosterone concentrations in eggs of

Corresponding author E-mail: mzeman@fns.uniba.sk

Japanese quail may change a behavioural phenotype towards "proactive" (Daisley *et al.*, 2005). Endogenous testosterone content in eggs of poultry species was studied only to a limited extent (Möstl *et al.*, 2001; Müller *et al.*, 2002; Hackl *et al.*, 2003). Possible variation related to age of hens, nutritional status, order of egg in laying sequence and other aspects is not known. These fluctuations may be partially responsible for physiological variability in performance of different flocks.

The present study focused on development of behavioural traits in Japanese quail offspring hatched from eggs with experimentally increased testosterone concentrations. Behavioural traits from hatching until the onset of sexual maturity were studied.

Material and methods

Japanese quail (*Coturnix japonica*) eggs were obtained from a breeding colony of the Institute of Animal Biochemistry and Genetics, SASci, Slovakia. Birds were of the same age (10 months) when eggs were collected. Eggs (mean weight in a range 9.9 – 10.2 g) were randomly assigned into two treatment groups, either the testosterone (51) or control (49) group. Each egg of the first group was immediately prior to incubation injected with 25 ng testosterone propionate (Agovirin, Léciva, Czech Republic) dissolved in 20 μl olive oil. This dose of testosterone was within the range of naturally occurring levels in Japanese quail eggs (Hackl *et al.*, 2003). Eggs of the control group received 20 μl of olive oil only. Injections were made through the blunt pole of the egg into the yolk. Prior to injection the eggshell was disinfected with 70 % ethanol and after it the hole was sealed with paraffin. Then all treated eggs were placed into a forced draught incubator (Bios Midi, Czech Republic), where they were automatic turned and incubated at the temperature 37 ± 0.2 °C and 50-60 % relative humidity.

After hatching, chicks from each treatment group were placed into the separated home cages and individually marked by colour rings on the leg. Hatchability of eggs did not differ between groups and was 44.5 % and 39.5 % for control and testosterone treated group, respectively. Quails were housed under the 24 hours light. Temperature was maintained at 35-37 °C during the first week then it was stepwise reduced to 21-24 °C in the third week. Feeding mash for young turkeys and water were provided *ad libitum*.

Only males (9 for control and 9 for testosterone treated group) were used in this experiment. Sex was definitively identified at the age of 3 weeks, when the sexually dimorphic coloration of breast feathers was apparent.

OPEN-FIELD TEST

Animals were tested at 2, 9, 23 and 37 days of age. They were individually transferred in a carton box into a testing room.

Open-field tests were performed in a wooden box with white painted walls and floor divided into 25 equal squares. It was situated in an area with temperature corresponding to the temperature in a breeding room. At 2 and 9 days of age birds were tested in the box measuring $25 \times 25 \times 25$ cm, while at 23 and 37 days of age its dimensions were $50 \times 50 \times 40$ cm (w x d x h). Tested quail were placed into the central square of the cage and their behaviours were recorded by videocamera for 5 minutes. The following behavioural categories were analysed from the videotapes: locomotor activity (number of lines crossed), vocalization (number of distress calls emitted), pecking (number of both wall- and floor-pecks), defecation (number of defecations), latency to explore (leave the start square) and latency to vocalize (first call).

DATA ANALYSIS

Following univariate analyses of normality and homogeneity all data were square root-transformed prior to statistical analyses. Behavioural data were analysed by the two-way ANOVA with treatment (testosterone and control) and age as factors.

Results

Open-field behaviours were influenced by both treatment and age. Two-way ANOVA of locomotor activity revealed significant effect of treatment ($F_{(1, 58)} = 5.396$, $p < 0.05$), age ($F_{(3, 58)} = 41.643$, $p < 0.001$) as well as interaction between the both factors ($F_{(3, 58)} = 4.36$, $p < 0.01$). There was an increased locomotor activity at 9 and 37 days of age in quails hatched from testosterone-treated eggs as compared to control birds (LSD tests, $p < 0.01$). Two days old quail chicks displayed higher locomotor activity compared to all other ages. In the control group the age-dependent decrease was stronger than in the testosterone-treated group (Fig. 1A).

The ANOVA proved the significant effect of treatment ($F_{(1, 58)} = 11.204$, $p < 0.01$), age ($F_{(3, 58)} = 76.335$, $p < 0.001$) and their interaction ($F_{(3, 58)} = 4.991$, $p < 0.01$) on open-field vocalization. At 9 days of age quails from the testosterone-treated group emitted an increased number of calls as compared to control quails (LSD test $p < 0.001$). Vocalization in the control group sharply decreased from day 2 to day 9 and then remained low until 37 days of age. In the testosterone-treated group the decrease in the number of calls was gradual over the age (Fig. 1B).

Significant differences in pecking were found both between treatment groups ($F_{(1, 58)} = 9.01$, $p < 0.01$) and between ages ($F_{(3, 58)} = 26.462$, $p < 0.001$) but not their interaction. This means that elevated testosterone levels in eggs induced in quails increased pecking as compared to the control group. Pecking activity within both groups decreased from hatching until 37 days of age and then it increased again (Fig. 1C).

The frequency of defecation was significantly affected by the treatment ($F_{(1, 58)} = 4.209$, $p < 0.05$) only. The mean number of defecations over the all ages in quails prenatally exposed to elevated testosterone was higher than in control birds (Fig. 1D).

Two-way ANOVA of the latency to explore (to leave the start square) revealed significant effect of treatment ($F_{(1, 58)} = 15.215$, $p < 0.001$) and age ($F_{(3, 58)} = 10.859$, $p < 0.001$). Interaction reached value $p = 0.078$ ($F_{(3, 58)} = 2.388$). At 9, 23 and 37 days of age latency to explore was reduced in the testosterone-treated group as compared to the control group (LSD tests $p < 0.05$). On day 2 after hatching quails of the control group showed significantly shorter exploration latency than at all other ages. There was a similar trend observed in testosterone-treated group but the differences between ages were not significant (Fig. 1E).

Two-way ANOVA revealed significant effect of treatment ($F_{(1, 58)} = 14.574$, $p < 0.001$), age ($F_{(3, 58)} = 31.009$, $p < 0.001$) and their interaction ($F_{(3, 58)} = 4.057$, $p < 0.05$) on latency to vocalize. At 9 and 23 days of age quails hatched from testosterone-treated eggs started to emit calls sooner than quails from control eggs (LSD tests, $p < 0.001$). In the control group latency to the first call showed a rapid increase from day 2 to day 9 and it remained high until day 37. Quails from the testosterone treated group displayed a gradual increase in latency to the first call over the age (Fig. 1F).

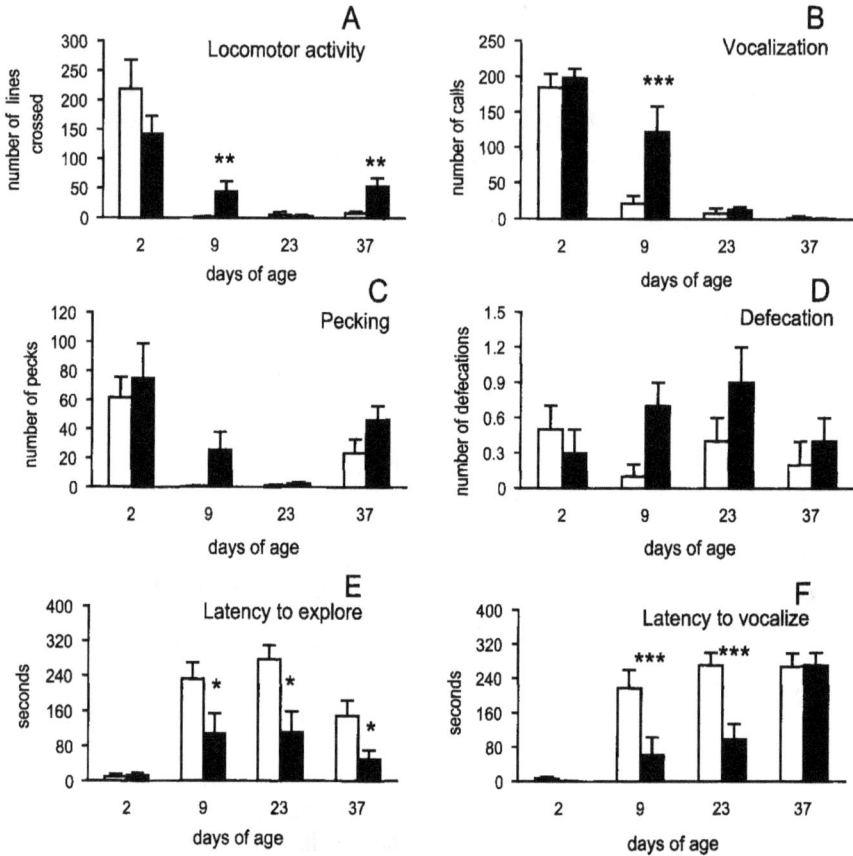

Figure 1. Open-field behaviours in control (empty bars) and testosterone-treated (filled bars) quails at four different ages. At the age of 2 and 9 days there were 8 and 7 birds in control and testosterone-treated group, respectively. At age 23 and 37 days nine birds in each group were tested. Values represent means ± SEM. Asterisks represent significance at the level * $p < 0.001$, ** $p < 0.01$, *** $p < 0.001$.

Discussion

Our results demonstrated that androgen exposure during embryonic development can influence in a different manner behavioural responses of Japanese quail offspring during ontogeny. While two-day-old quail chicks hatched from testosterone-treated eggs do not differ in their behaviours in the open-field from controls, apparent behavioural differences between both groups were observed on day 9 after hatching. Birds exposed to elevated testosterone in the egg displayed higher levels of locomotion, vocalization and pecking that were accompanied by both shorter latency to vocalize and shorter latency to explore in comparison with control quails. This pattern was observed in some behavioural parameters also in subsequent ages and higher locomotor activity as well as shorter latency to explore was found at 37 days of age.

It is generally accepted that behavioural responses in the open field are under the control of both social motivation and effort to avoid a potential predator (Faure *et al.*, 1983). While common responses in a novel environment are associated with a degree of fear, longer freezing and both suppressed vocalization and motor behaviour represent higher level of fearfulness, indicating higher emotionality (Ginsburg *et al.*, 1974). Less inhibition of behaviour expressed by birds from testosterone treated eggs in comparison with controls, observed in our study, suggests their lower fear responses to challenge. Indeed, this interpretation is comparable to higher exploratory pecking behaviour and shorter latency for the first step displayed by quail chicks with lower general emotional reactivity (Richard-Yris *et al.*, 2005). Moreover, higher defecation frequency was recorded in the prenatally testosterone treated group as compared to control one in our study. This observation is not consistent with an expected negative correlation between defecation and open-field ambulation described in classical studies. The possibility of dissociation between open-field ambulation and defecation has been questioned in a review by Ramos and Mormede (1998). Moreover, positive correlation between defecation and number of lines crossed in an open field and defecation suppression during intense fear has been reported in chicks (Marin *et al.* 1997).

Several studies showed that both vocalization and locomotor activity are androgen-dependent behaviours in Japanese quail (Wada, 1982) and that testosterone is able to induce some morphological changes in steroid-sensitive areas of the non-songbird midbrain and the syrinx, structures responsible for vocalization (Beani *et al.*, 1995). Such consequences resulted from post hatching androgen treatment and thus represent a response to higher levels of these steroid hormones in circulation. On the other hand, embryonic androgen exposure is likely to be involved in organizational pathways that have a strong impact during development of the central nervous system. Possible sites for action of maternal testosterone have been suggested by presence of functional androgen receptors in the motor nuclei of the embryonic zebra finch hindbrain (Godsave *et al.*, 2002). Whether yolk testosterone affects directly motor systems or modifies brain sensitivity by another mechanism, it is probable that the existence of these individual differences at the level of the neural networks results in individual differences in the behavioural expression (Koolhaas *et al.*, 1997).

Experimental manipulations of yolk androgen concentrations in avian eggs elicit predominantly positive effects on offspring early postnatal life (Schwabl, 1996; Eising and Groothuis, 2003), although it appears that this relationship is species-specific (Sockman and Schwabl, 2000). We observed behavioural differences in locomotor activity between testosterone treated and control birds at the end of experiment (37 days) that is around the age of sexual maturation in Japanese quail. This finding extends available data on long-lasting consequences of maternal androgens, which were shown to organize some morphological and behavioural traits indicating fitness and reproductive success in various bird species (Strasser and Schwabl, 2004; Uller *et al.*, 2005). Endogenous variation of maternally deposited hormones, including testosterone, may represent one source of variation in performance of poultry species. In altricial birds differences in yolk hormone content are related to social experience of the mother (Gil *et al.*, 1999; Whittingham and Schwabl, 2002; Pilz and Smith, 2004). In poultry species we need more data about this relationship to consider how age of hens, nutritional status, health conditions etc. may contribute to yolk testosterone variation and further postembryonic development and behavioural phenotype of offspring.

Our data suggest that increased yolk testosterone produces less fearful Japanese quail offspring. This effect seems to be age-dependent. Lowered fearfulness is in accordance with similar findings of Daisley *et al.* (2005) in the same species based on the series of behavioural tests. However there are some discrepancies in the open-field behaviour between both studies. Daisley *et al.* (2005) found that birds from testosterone treated eggs called less and their latency to vocalize was longer as compared to control group. These differences can be explained by different ages of tested birds. We exposed them to the open-field repeatedly during development while Daisley *et al.* (2005) exposed them during the first week of life. This period coincides with sensitive phase when stable fear-related responses do not appear to be fully formed yet and also they can be affected by early experiences (Heiblum *et al.*, 1998). In terms of phenotypic plasticity it may be advantageous to change behavioural strategies during ontogeny depending on given environmental conditions (Price *et al.*, 2003).

Since our data include only males, we cannot consider sex-specific effects of maternal testosterone in avian eggs. However a recent study in the same species demonstrated that early androgen exposure influences individual behavioural phenotype equally in both males and females (Daisley *et al.*, 2005). On the other hand opposite results were found in zebra finches, suggesting sex-specific effects of increased maternal testosterone on begging behaviour and growth (von Engelhardt *et al.*, 2004).

In conclusion, our data suggest an involvement of yolk testosterone in the organizing of individual differences in emotional behaviour of precocial Japanese quail. Effects of testosterone seem to vary during ontogeny that enables better coping with prevailing environmental conditions. During a short period after hatching, when hatchlings have limited physical abilities to cope actively with challenges, they exhibited increased fearfulness and behaved in a passive manner in an unknown environment. When their physical capabilities increased, starting from day 9, birds hatched from testosterone treated eggs changed their strategy and reacted to stimuli from surrounding environment in a "proactive" manner.

Acknowledgement

This study was supported by the grant VEGA No. 1/1294/04.

References

BEANI, L., PANZICA, G., BRIGANTI, F., PERSICHELLA, P., and DESSI-FULGHERI, F. (1995) Testosterone-induced changes of call structure, midbrain and syrinx anatomy in partridges. *Physiology and Behavior* **58**: 1149-57.

DAISLEY, J.N., BROMUNDT, V., MÖSTL, E., and KOTRSCHAL, K. (2005) Enhanced yolk testosterone influences behavioral phenotype independent of sex in Japanese quail chicks *Coturnix japonica. Hormones and Behavior* **47**: 185-194.

EISING, C.M., and GROOTHUIS, T.G.G. (2003) Yolk androgens and begging behaviour in black-headed gull chicks: an experimental field study. *Animal Behaviour* **66**: 1027-1034.

FAURE, J.M., JONES, R.B., and BESSEI, W. (1983) Fear and social motivation as factors in open-field behaviour of the domestic chick: a theoretical consideration. *Biology of Behaviour* **8**: 103-116.

GIL, D., GRAVES, J., HAZON, N., and WELLS, A. (1999) Male attractiveness and differential testosterone investment in zebra finch eggs. *Science* **286**: 126–128.

GINSBURG, H.J., BRAUD, W.G., and TAYLOR, R.D. (1974) Inhibition of distress vocalizations in the open field as a function of heightened fear or arousal in domestic fowl (*Gallus gallus*). *Animal Behaviour* **22**: 745-749.

GODSAVE, S.F., LOHMANN, R., VLOET, R.P.M., and GAHR, M. (2002) Androgen receptors in the embryonic zebra finch hindbrain suggest a function for maternal androgens in perihatching survival. *Journal of Comparative Neurology* **453**: 57-70.

GROOTHIUS, T.G.G., MULLER, W., VON ENGELHARDT, N., CARERE, C., and EISING, C.M. (2005) Maternal hormones as a tool to adjust offspring phenotype in avian species. *Neuroscience and Biobehavioral Reviews* **29**: 329-352.

HEIBLUM, R., AIZENSTEIN, O., GVARYAHU, G., VOET, H., ROBINZON, B., and SNAPIR, N. (1998) Tonic immobility and open field responses in domestic fowl chicks during the first week of life. Applied Animal Behaviour Science **60**: 347-357.

HACKL, R., BROMUNDT, V., DAISLEY, J., KOTRSCHAL, K., and MÖSTL, E. (2003) Distribution and origin of steroid hormones in the yolk of Japanese quail eggs (*Coturnix coturnix japonica*). *Journal of Comparative Physiology B* **173**: 327-31.

KOOLHAAS, J.M., DE BOER, S.F., and BOHUS, B. (1997) Motivational systems or motivational states: Behavioural and physiological evidence. *Applied Animal Behavioural Science* **53**: 131-143.

MARIN, R.H., MARTIJENA, I.D., and ARCE, A. (1997) Effect of diazepam and A beta-carboline on open-field and T-maze behaviors in 2-day-old chicks. *Pharmacology Biochemistry and Behavior* **58**: 915-921.

MÖSTL, E., SPENDIER, H., and KOTRSCHAL, K. (2001) Concentration of immunoreactive progesterone and androgens in the yolk of hens´ eggs (*Gallus domesticus*). *Weiner Tierärztliche Monatsschrift* **88**: 62-65.

MÜLLER, W., EISING, C.M., DIJKSTRA, C., and GROOTHUIS, T.G. (2002) Sex differences in yolk hormones depend on maternal social status in Leghorn chickens (*Gallus gallus domesticus*). *Proceedings of the Royal Society of London. Series B, Biological Sciences* **269**: 2249-55.

PILZ, K.M., and SMITH, H.G. (2004) Egg yolk androgen levels increase with breeding density in the European starling, *Sturnus vulgaris*. *Functional Ecology* 18: 58-66.

PRICE, T.D., QVARNSTROM, A., and IRWIN, D.E. (2003) The role of phenotypic plasticity in driving genetic evolution. *Proceedings of the Royal Society of London. Series B, Biological Sciences* **270**: 1433-40.

RAMOS, A., and MORMEDE, P. (1998) Stress and emotionality: a multidimensional and genetic approach. *Neuroscience and Biobehavioral Reviews* **22**: 33-57.

RICHARD-YRIS, M.A., MICHEL, N., and BERTIN, A. (2005) Nongenomic inheritance of emotional reactivity in Japanese quail. *Developmental Psychobiology* **46**: 1-12.

SCHWABL, H. (1996) Maternal testosterone in the avian egg enhances postnatal growth. *Comparative Biochemistry and Physiology* **114**: 271-276.

SOCKMAN, K.W., and SCHWABL, H. (2000) Yolk androgens reduce offspring survival. *Proceedings of the Royal Society of London. Series B, Biological Sciences* **267**: 1451-1456.

STRASSER, R., and SCHWABL, H. (2004) Yolk testosterone organizes behavior and plumage coloration in house sparrows (*Passer domesticus*). *Behavioral Ecology and Sociobiology* **56**: 491-497.

ULLER, T., EKLÖF, J., and ANDERSSON, S. (2005) Female egg investment in relation to male sexual traits and the potential for transgenerational effects in sexual selection. *Behavioral Ecology and Sociobiology* **57**: 584-590.

VON ENGELHARDT, N., CARERE, C., DIJKSTRA, C., and GROOTHUIS, T.G.G. (2004) Elevation of yolk testosterone abolishes sex differences in begging and growth of zebra finches. In: VON ENGELHARDT, N. (ed.) *Proximate control of avian sex alloca-tion, a study in zebra finches. PhD Thesis* University of Groningen, Research Unit Ani-mal Behaviour, NL, pp. 77-89.

WADA, M. (1982) Effects of sex steroids on calling, locomotor activity, and sexual behav-ior in castrated male Japanese quail. *Hormones and Behavior* **16**: 147-157.

WHITTINGHAM. L.A., and SCHWABL, H. (2002) Maternal testosterone in tree swal-low eggs varies with female aggression. *Animal Behaviour* **63**: 63-67.

Permeation of volatile chemical compounds through an eggshell as determined by a newly developed high-sensitivity analytical system

[1]AKITO SHIMOUCHI, [2]YASUHIRO CHIBA, [1]JAMES T PEARSON, [1]MIKIYASU SHIRAI, [2]HIROSHI TAZAWA

[1]National Cardiovascular Center Research Institute, Osaka, Japan; [2]Muroran Institute of Technology, Hokkaido, Japan

For chicken embryos, numerous studies have reported the exchange of water vapour and respiratory gases such as oxygen and carbon dioxide; however, little is known whether, and to what extent, chemical permeation through an eggshell occurs in terms of volatile chemical compounds. In this study, we determined very low levels of volatile chemical compounds released from developing chicken embryos, using a new biogas-sampling and analysis system incorporating an Atmospheric Pressure Ionization Mass Spectrometer with high sensitivity.

White Leghorn eggs were incubated for 0, 5, 10, 15 and 20 days in a commercially available incubator. On the day of experiments, to measure volatile compounds released from the eggshell surface, air cell of each egg was positioned upwards in a thermo-regulated chamber continuously ventilated by highly purified air as a carrier gas. First, the mass spectrum of volatile chemical compounds released from the egg was determined. Thereafter, the eggshell over the air cell was partially removed and its mass spectrum was determined again.

The ion intensity (I.I.) of the gases released from the shell-free egg over the air cell was remarkably increased compared with the intact egg on any incubation days. On the assumption that I.I. has a linear relationship with the concentration at each mass to charge ratio (m/z), the permeation ratio of volatile compounds through the eggshell at each m/z number was defined as a ratio: I.I. intact / I.I. eggshell removed. The calculated permeation ratio at each m/z number was decreased in proportion. The permeation ratio of most molecules did not exhibit significant changes during embryonic development. These results suggest that permeation of volatile chemicals through the eggshell may depend upon the physicochemical characteristics of volatile chemical compounds and that the eggshell may play a role in preventing the loss of larger volatile molecules.

Keywords: eggshell, chick embryo, volatile chemical compounds, gas exchange, mass spectrometry

*Corresponding author: Akito Shimouchi
E-mail: ashimouc@res.ncvc.go.jp

Introduction

For chicken embryos, numerous studies have reported on the exchange of water vapour (Ar *et al*, 1974), hydrogen, nitrogen and such respiratory gases as oxygen and carbon dioxide (Romanoff,1943; Lomholt, 1976). In particular, the diffusion of oxygen and carbon dioxide across eggshells was extensively investigated mainly because gas exchange across the eggshell is a basic and simple model of respiration (Rahn and Paganelli, 1990).

In the developing embryo inside an eggshell, numerous metabolic processes are occurring whether they are developing normally or not. Among the products, volatile chemical compounds at incubation temperature are produced and gaseous compounds in the chick embryo diffuse across the inner shell membrane, outer shell membrane and porous eggshell. Between the former two membranes, air space is formed at the blunt end of the egg. This air space contains numerous gaseous compounds diffused not only from inside the internal membrane but from the outside the outer membrane and eggshell pores, possibly depending upon their incubation periods (Romijn *et al*.1938) and microclimate (Ar *et al*, 2001). Curiously, little is known about whether, and to what extent, chemical permeation through an eggshell occurs in terms of volatile chemical compounds (Romanoff 1943; Rahn *et al*, 1987; 1990), regardless of detections of numerous volatile chemicals inside the non-fertile eggshell (MacLeod *et al*, 1976: Bennett *et al*, 1978; Gil *et al*, 1981; Stein *et al*, 1990; Vannelswyk *et al*, 1995; Cherian *et al*, 2002).

In this study, we first determined that very low levels of volatile chemical compounds are released from developing chicken embryos, using a new biogas-sampling and analysis system incorporating an Atmospheric Pressure Ionization Mass Spectrometer (APIMS) with high sensitivity. We next compared the very low levels of volatile chemical compounds released from intact or eggshell-removed eggs and considered the permeation of volatile chemicals across the eggshell and outer membrane of fertile chicken eggs during development.

Methods

EXPERIMENTAL SETUP

In this study, we utilized an APIMS (Nippon API, Ltd., Co., Tokyo, Japan), which can detect ppm or lower levels of gaseous compounds (Lovett *et al*, 1979). To introduce chemical compounds released from the eggs, we developed a new gas-trap system in which an air cell of each egg was positioned upwards to analyze gas that may contain numerous volatile compounds. This system consisted mainly of mass-flow controllers, molecular sieves as gas cleaners and a thermo-controlled chamber through which the cleaned gas passes and volatile chemicals released from the egg were sampled. The chamber was columnar with a volume of 337ml (radius 80mm, height 67mm). The internal surface of the chamber, gas lines and connectors were made up of electrolytic-polished SUS316 to keep their internal surfaces clean. The inside space of the chamber and the artificial air were maintained at 38°C by thermo-regulators. Highly purified artificial air with 21% oxygen balanced with nitrogen (Taiyo-Toyo Sanso Ltd. Co. Tokyo, Japan) was introduced into the chamber using a mass-flow controller at a flow rate of 100 ml/min. While maintaining a constant mixing ratio, 20 ml/min of outflow gas from the chamber was diluted with the same purified gas and passively introduced into the APIMS. The APIMS was set as follows: ion chamber temperature = 120°C discharge current = 10 µA, SEM voltage = 1600V, Drift voltage = 40V, Q-Pole lens operation: Scan mode (mass to charge ratio (m/ z): 3 -

150, 45 s / scan), Detector polarity: Negative, Detector sensitivity: 1.0×10^{-9} and 1.0×10^{-7}A. Each scan speed was 90sec. The experimental setup is shown in Fig. 1.

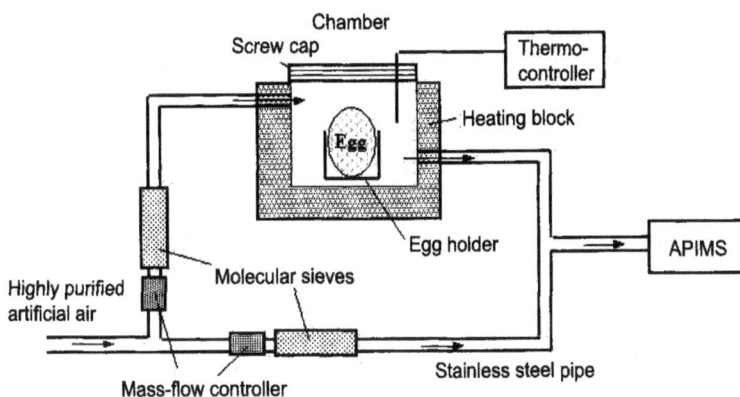

Figure 1. Biogas sampling system for high sensitive analysis using an Atmospheric Ionization Mass Spectrometer (APIMS).

EXPERIMENTAL PROTOCOLS

Fertile White Leghorn eggs were purchased from a local supplier. Twenty-four eggs were randomly divided and placed at two separate positions in the chamber with or without turning, the latter condition of which was to cause the increase of mortality or immature growth of the eggs. Twelve eggs were turned normally every 4 hours in a commercially available incubator at 38°C and 55% humidity, while the remaining eggs were not turned in the same incubator. Eggs in each group were sampled for the following experiments after incubation of 0, 5, 10, 15 and 20 days.

Among the twelve eggs undergoing normal turning, 11 eggs exhibited normal development and were categorized as the survival group. The remaining egg did not have a visible embryo after 15 days' incubation and seemed to be infertile. This egg was included in the non-survival group. Among the other twelve eggs without turning, three eggs had normal embryo at 15 days, which were counted in the survival group. One egg with a normal embryo at 5 days was smashed during the experimental procedure. The other eight eggs had no visible embryo or a dead embryo with poor development, and were grouped into the non-survival group. We therefore obtained 14 survival eggs with normal development and 9 non-survivals with poor development or without a visible embryo. These 23 eggs were used to measure chemical compounds released from the egg before and after removal of the eggshell and outer membrane on the air cell.

On the day of experiments, to measure volatile compounds released from the eggshell surface, each egg was placed vertically so as to stand air cell upwards in a thermo-regulated chamber continuously ventilated with highly purified air as a carrier gas. First, the chamber was circulated with the artificial air and the background API mass spectrum was monitored for 15 min. Then, an egg sampled from the incubator was put in the chamber in which the air cell was positioned upwards by using a holder. The volatile chemical compounds released from the egg in the closed chamber were monitored for 30 min. Thereafter, the eggshell over the air cell was partially removed and its mass spectrum was measured again for 30 min. The scanning cycle to obtain the mass spectrum from m/z = 3-150 was 90 sec.

DEFINITION OF PERMEATION RATIO

As it was impossible to survey all relationships between the concentration and ion intensity at each m/z number, we assumed that both parameters, at any given m/z number and in the observed concentration range, had each linear correlation and that the permeation ratio of volatile compounds through the eggshell at each m/z number could be defined as the ratio: I.I.intact / I.I.eggshell removed.

Results

An example of the mass spectrum of chicken eggs, which was a survival case at 15 days' incubation, is shown in Fig. 2. In the negative charge mode, the background spectrum in the chamber without an egg indicated very low ion intensity except at m/z = 16, 32 and 48, which were oxygen anions such as O^-, O_2^- and O_3^-, respectively, produced in the reaction chamber of APIMS (Fig. 2-a). When an egg was placed in the chamber, volatile chemical compounds were detected at around m/z = 78, 95, 109 and 124 in addition to m/z = 16, 32, 48 as mentioned above (Fig. 2-b). After removal of the eggshell with the outer membrane, inside of which formed the air cell, the ion intensity at m/z = 60 or more was remarkably increased (Fig. 2-c). This mass spectrum pattern was seen regardless of survival or non-survival (data not shown).

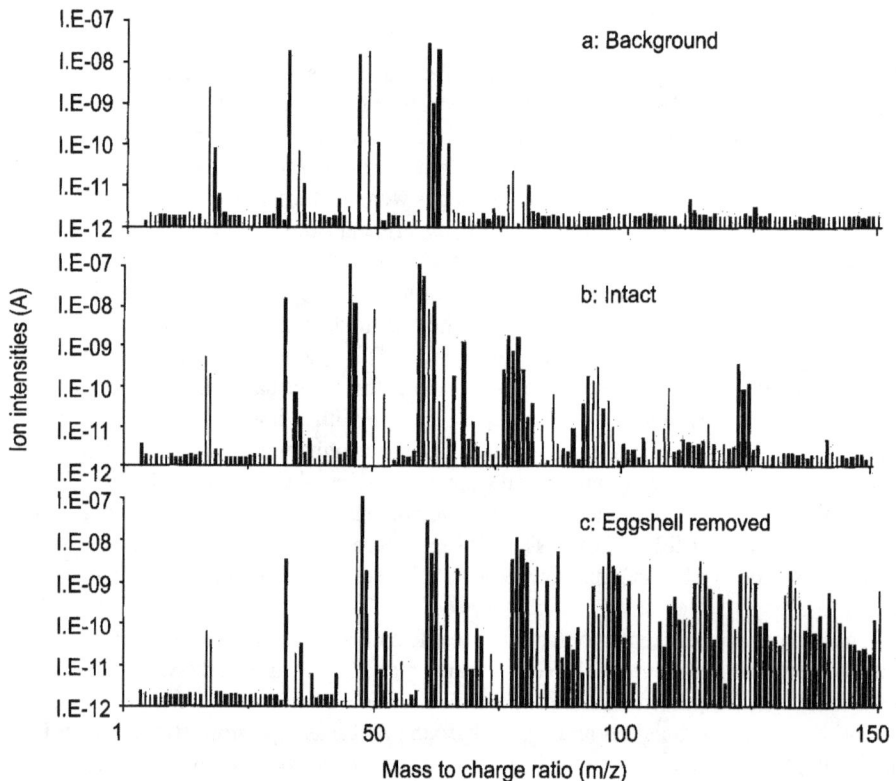

Figure 2. An example case of mass spectra obtained from a: background of the biogas sampling system, b: from the surface of an egg in the chamber and c: shell-removed egg in the air cell.

In Fig. 3, the average permeation ratio on each incubation day in the survival and non-survival groups is illustrated, because no significant differences were detected among m/z dependencies of the permeation ratio on each incubation day and between survival and non-survival groups. The calculated permeation ratio at each m/z number tended to decrease in proportion to the increase of m/z as shown in Fig. 3.

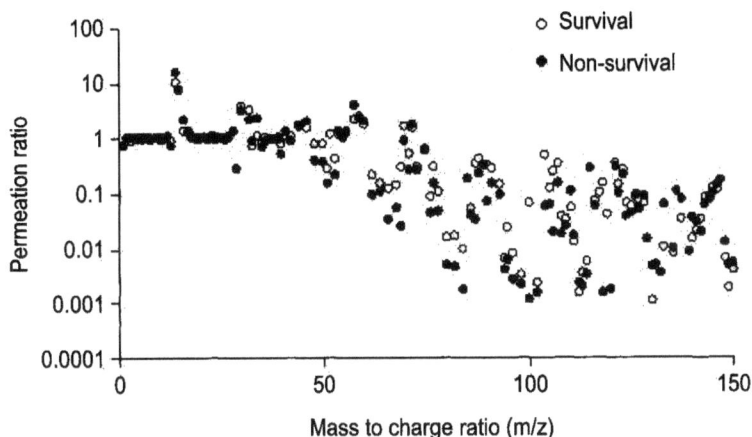

Figure 3. Dependence of permeation ratio on the developmental conditions of eggs (survival vs non-survival). Permeation ratio at each m/z number is the average of 5 to 20 days' incubation.

Discussion

In this study, we developed a new biogas-sampling and analysis system incorporating an APIMS with high sensitivity. Using this system, we found that several volatile chemical compounds were released from the surface of intact eggs and that many more chemicals were released from the surface of the internal membrane, i.e., after removal of the eggshell and outer membrane of the air cell.

Several investigators especially in the fields of food chemistry (MacLeod *et al*, 1976: Bennett *et al*, 1978; Gil *et al*, 1981; Stein *et al*, 1990; Vannelswyk *et al*, 1995; Cherian *et al*, 2002) reported up to 96 kinds of peaks, i.e. volatile chemical compounds inside the eggshell of non-fertile eggs. Most of them were organic chemical compounds detected mainly by gas chromatograph and/or gas chromatograph - mass spectrograph with standard libraries. In this study, mass spectra of chemicals released from the surface of fertile eggs with or without an eggshell at the air cell were obtained, using a negative APIMS mode as a detector. The ability of chemicals to have a negative charge depends upon the physicochemical characteristics. Well-known negatively charged chemicals are $R-SO_3X$, $R-O-SO_3X$, $R-COOX$, C_6H_6-OH, $R_1-COCH-R_2R_3$, $R_1-COCH=CHCH-R_2R_3$, $CN-CH_2-CN$, $R-OH$ and so on, where R is a residue. Therefore, these chemicals might be detected in these mass spectra by the negative mode of APIMS. Because available libraries for APIMS have unfortunately not been established, each chemical representing the ion intensities were not identified. However, our results indicate that high intensity m/z numbers in the spectrum of released gas from the surface of intact eggs were limited in case of whole eggs, while shell-removed eggs exhibited much higher intensity at nearly all m/z numbers. These results indicate that numerous kinds of volatile chemicals are released from the surface of the inner membrane inside the eggshell or across the inner membrane, and suggest that the eggshell works as diffusion barrier.

Although ion intensity does not always have a linear relationship to the chemical concentration with any given m/z number, on the assumption that ion intensity has a linear relationship with concentrations at any given m/z number, the permeation ratio of chemical compounds across the egg shell and outer membrane at the air cell can be simply defined as a ratio: I.I.$_{intact}$ / I.I.$_{eggshell\ removed}$ to distinguish from originally defined permeability (Romanoff, 1942; Lomholt 1976; Kayer *et al*, 1981; Wangensteen *et al*, 1970-1971, Kuchai *et al*, 1971). If the chemicals are released from the contaminations and/or from the cuticle that covers the surface of the egg shell, the ion intensity after removal of the egg shell will be decreased and therefore the permeation ratio will be higher. In the ion chamber of APIMS, oxygen anions play the role of electron donor to the other chemicals introduced. The total number of electrons in the reaction chamber is limited. When numerous compounds capable of being negatively charged are carried into the chamber, the remaining oxygen anion that exhibits high intensity will be decreased because of electron donation to the other chemicals. Therefore, the permeation ratio of oxygen anions and the related chemicals at around m/z = 16, 32, 48 is overestimated, which is not actual oxygen permeation ratio across the eggshell. Also in case of nitrogen and its related chemicals, caution is needed to consider the implications of the permeation ratio spectrum because nitrogen is already present in the carrier gas.

Alternatively, if the eggshell is a diffusion barrier for the volatile chemical compounds, the chemical compounds released from the shell-removed eggs will be increased, and thereby the permeation ratio will be lowered. Even if the electron-donor effect of the oxygen anion is excluded, our results still seem to suggest that the permeation ratio is gradually decreased in proportion to m/z numbers, which, although scattered, also suggest that permeation ratio depends upon chemical species. We speculated that decreases of permeation ratios in proportion to m/z could be partially attributed to the molecular size, because it is generally accepted that diffusion coefficient decreases with increase of molecular weights. However, for example, carbohydrate with m/z=100 is nano-order size, while pore size of the chicken eggshell is 10 micron-order. Therefore, the pore to molecular size ratio is calculated to 1000 to 10,000, which is considerably large. Therefore, anatomical properties of egg-shell pore and/or physicochemical properties of the internal surface of the pores must be also taken into consideration (Board *et al*, 1991).

Before performing this study, we speculated that permeation ratio at each m/z numbers would be increased during developments in survival groups, because the calcium mobilization from eggshell to the embryo (Simkiss, 1991) may affect porosity of the eggshell, thereby increasing the permeation ratios. In most cases of m/z numbers, permeation ratios both in survival and non-survival groups were not significantly changed during development. We therefore summarized the data as shown in Fig. 3. These results indicate that the permeation ratios were not dramatically affected by the development of embryo regardless of their survival.

In summary, we developed a new type of gas sampling system for the analysis of biogas from the surface of avian eggs, using an Ambient Pressure Ionization Mass Spectrometer with high sensitivity. This study revealed that numerous volatile compounds were partially released from the surface of eggshells, which act as a diffusion barrier for larger molecules, and may prevent the loss of volatile organic compounds from inside the eggshell. Further studies should identify the chemicals released from eggs, origin of the chemicals, and specific compounds exhibiting changes of permeation ratio during development, because they may clarify the mechanism of diffusion across the porous eggshell in more detail.

Acknowledgements

This study was supported by Grants in Aid from the Japan Society for the Promotion of Science (JSPS) and the New Energy and Industrial Development Organization (NEDO) of Japan.

References

AR, A. & SIDIS, Y. (2001). Nest microclimate during incubation. In: *Avian Incubation: Behaviour, environment and evolution*: 143-160. Deeming, D. C. (Ed.). Oxford, U.K.: Oxford University Press.

Ar. ACV, Paganelli CV, Reeves RB, Greene DG, Rahn H. (1974) The avian egg:water vapour conductance, shell thickness, and functional pore area. *The Condor* 76:153-158.

Bennett SJ, Britton WM, HORVAT R J. (1978) Gas-chromatographic analysis of egg volatiles. *Poultry Science* 57: 1118.

Board RG, Sparks NHC. (1991) Shell structure and formation in avian eggs. In: *Egg incubation: its effects on embryonic development in birds and reptiles*: 71-86.Deeming DC, Ferguson MWJ (Ed.). Cambridge, U.K.: Cambridge University Press.

Cherian G, Goeger MP, Ahn DU. (2002) Dietary conjugated linoleic acid with fish oil alters yolk n-3 and trans fatty acid content and volatile compounds in raw, cooked, and irradiated eggs. *Poultry Science* 81: 1571-1577.

Gil V, MacLeod AJ. (1981) Synthesis and assessment of 3 compounds suspected as egg aroma volatiles. *J Agri Food Chem* 29: 484-488.

Kayar SR., Snyder GK, Birchard GF, Black CP. (1981) Oxygen permeability of the shell and membranes of chicken eggs during development. *Respir Physiol* 46:209-221.

Kuchai H, Steen JB. (1971) Permeability of the shell and shell membrane of hen's eggs during development. *Respir Physiol* 11:265-278.

Lomholt JP. (1976) The development of the oxygen permeability of the avian egg shell and its membranes during incubation. *J Exp Zool* 198: 177-184.

Lovett AM, Reid NM, Buckley JA, French JB, Cameron DM. (1979) Real-time analysis of breath using an atmospheric pressure ionization mass spectrometer. *Biomed Mass Spectrom* 6(3): 91-7.

MacLeod AJ, Cave SJ. (1976) Variations in volatile flavor components of eggs. *J Sci Food Agri.* 27: 799-806.

Rahn H, Paganelli CV, Ar A. (1987) Pores and gas exchange of avian eggs: a review. *J Exp Zoo* suppl. I:165-172.

Rahn H, Paganelli CV. (1990) Gas fluxes in avian eggs: driving forces and the pathway for exchange. Comp. *Biochem Physiol* 95A(1): 1-15.

Romanoff AL. (1943). Study of various factors affecting permeability of birds' eggshell. *Fed. Res.* 8: 212-223.

Romijn, C. & Roos, J. (1938). The air space of the hen's egg and its changes during the period of incubation. *J. Physiol.* **94**: 365-379.

Simkiss K. (1991) Fluxes during embryogenesis. In: *Egg incubation: its effects on embryonic development in birds and reptiles*: 71-86.Deeming DC, Ferguson MWJ (Ed.). Cambridge, U.K.: Cambridge University Press.

Stein VB, Narang RS. (1990) A simplified method for the determination of volatiles in eggs using headspace analysis with a photoionization detector. *Arch Environ Contam Toxicol* 19: 593-596.

Vanelswyk ME, Dawson PL, Sams AR. (1995) Dietary menhaden oil influences sensory characteristics and headspace volatiles of shell eggs. *J food sci* 60: 85-89.

Wangensteen OD, Wilson D, Rahn H. (1970-71) Diffusion of gases across the shell of the hen's egg. *Respir Physiol* 11(1):16-30.

Effects of artificial insemination as a handling stress on egg weight, yolk corticosterone content and embryonic mortality (Pilot study)

SZÓKE, ZS, VÉGI, B., VARGA, Á., LENNERT, L, PÉCZELY P.* AND BARNA J.

Dept. of. Avian Reproduction, Institute for Small Animal Research 2100 Gödöllő POB.:417. Hungary, *Dept. of Reproductive Biology, Szent István University, 2100 Gödöllő, Páter K. u. 1

The aim of the study was to determine the changes in egg size, concentrations of corticosterone in egg yolks and in the rate of embryonic death following artificial insemination (AI) as a stressor, in the domestic hen.

Fourteen hens of the experimental (AI) group were inseminated with fresh, diluted semen (300 million spermatozoa) 3 times a week, while as a control, 14 hens and 2 cockerels were mated naturally without disturbances. During 8 weeks eggs were collected and weighed then incubated for 5 days. After determination of embryonic status (live, dead, developmental stages) yolk samples were collected for analyzing of corticosterone content by RIA.

As a result of AI the average weight of eggs increased by 1 gram, however, the eggs containing abnormal embryos increased by 4-5 grams. Thus, a relationship could be found between the large-size eggs and dead or abnormal embryos. Parallel with the egg-weight, artificial insemination significantly increased the glucocorticoid level of the yolks. Moreover, in the eggs with abnormal or dead embryo the corticosterone level was even higher than in the eggs with a normal, live embryo. In the AI group the number of abnormal and dead embryos increased by 15-23 % on the first 5 weeks, then it decreased. On the 6-7th weeks the embryos in all eggs were alive and showed normal development.

It is generally known that the weights of eggs coming from birds kept in cages and reproduced by AI are usually higher compared to naturally mating birds kept on deep litter. While this study could support this observation, however, it proved that the embryonic mortality is correlated with corticosterone level during the first five weeks of the treatment. According to the study, 6 weeks are needed for the adaptation of hens to the AI procedure.

Keywords: embryonic development, corticosterone, artificial insemination, domestic fowl

*Corresponding author
E-mail: szoke@katki.hu

Introduction

Females influence the phenotype of their offspring not only by genetic means , but also through direct "maternal effects". These so-called "maternal effects" mean that environmental conditions experienced by the parents influence the offspring phenotype, potentially enhancing offspring fitness (Mousseau and Fox, 1998). Genetic resources, in the form of DNA, guide the offspring's development and limit its phenotypic potential. Non-genetic resources such as mitochondria, RNA, proteins, lipids, water, antibodies, hormones, are also essential to the development of embryos in all animal species, including the birds as well.

Investigation of steroids in avian egg yolk dates back to the 1920s. Early bioassays used to detect steroids in hens'eggs were inconclusive regarding their presence (Hertelendy and Common, 1965). In 1942, Riddle and Dunham discovered that the chicks hatching from oestradiol treated hens (*Gallus domesticus*) were feminised, similarly to if the eggs had been directly injected. Following these results, Arcos (1972) demonstrated that injection of radioactive oestradiol (E2) into laying hens resulted in transfer of radioactive oestrogens - including oestradiol and oestrone - to yolks of subsequently laid eggs.

In the last decade Schwabl (1993) reintroduced the question of the importance of steroid hormones in avian eggs. He showed that canary and zebra finch egg yolks contain high levels of androgens (testosterone, dihydrotestosterone, and androstendion), compared to low levels of oestradiol and very low levels of corticosterone (B) in zebra finch egg yolk. In canary eggs corticosterone wasn't detectable. Following these results more hormones have been detected in avian eggs: several type of thyroid hormones (McNabb and Wilson, 1997), progesterone (Lipar *et al*, 1999) and corticosterone (Hayward and Wingfield, 2004).

Many factors may elevate the concentrations of plasma glucocorticoids in vertebrates. In mammals the increased glucocorticoid levels are known to depress the intrauterine development of embryos. Ward (1972) found that maternal stress in pregnant rats feminized male offspring, decreased the fertility and fecundity of female offspring (Herrenkohl, 1979), reduced learning ability and increased the response of the hypothalamo-adrenocortical axis (Takahashi *et al*, 1992).

In egg-laying vertebrates, embryos are exposed only to those maternal hormones that were deposited into the yolk during the relatively short period of yolk formation (Hayward and Wingfield, 2004). Most yolk is incorporated into the follicle during the last 72 hours before ovulation (Romanoff and Romanoff, 1949). Hayward and Wingfield (2004) implanted corticosterone-filled or empty silastic capsules into female Japanese quail and measured concentrations of corticosterone in the yolk. They found that the corticosterone implants significantly increased the corticosterone level of yolk. Furthermore chicks of corticosterone-implanted mothers hatched worst, grew more slowly than controls and showed increased activity of the hypothalamo-adrenal axis in response to capture and restraint as adults.

In fish it has been proved that yolk cortisol level has been associated with reduced length of larvae at hatching (McCormick, 1999), increased the incidence of abnormal larvae (Morgen *et al*, 1999) and egg mortality (Pottinger and Carrick, 2000.).

In mallard, the faecal B concentration of mothers increased, and an increased quantity of B was deposited into the yolk on the effects of handling stress. The elevated B concentration of the yolk showed positive correlation with the increasing egg weight (Szőke *et al*, 2004)

There are only a few data regarding egg size changes during the production period. It is known that the size and composition of the egg depends on the hens' nutritional status, body weight and age, their genetic make up and the lighting regime for the flock (Etches, 1995).

Sinervo and DeNardo, (1996) found that in side-blotched lizards the chronically elevated plasma corticosterone level in the female was accompanied by an increase in egg size as measured by egg mass and independent of any change in clutch size.

According to several poultry breeders it is a generally accepted fact that the weights of eggs coming from breeds kept in cages and reproduced by AI are usually higher compared to naturally mating breeds kept on deep litter.

The above mentioned results regarding the yolk hormonal status support the idea that increased plasma corticosterone level in the female birds can appear in the egg yolk. Knowing that the plasma corticosterone level is a sensitive indicator of the environmental stress factors, in the present study the effects of the artificial insemination - as a stressor - on the yolk corticosterone level, the egg weight, and the embryonic development of domestic hen were investigated.

Material and methods

ANIMALS, AI PROCEDURES

Twenty-eight New Hampshire hens were randomly divided into two groups: control and experimental (AI) group. The two groups of females were kept in the same conditions: in aviaries on deep-litter, under 13L-11D lighting program and *ad libitum* access to food and water. Before the investigations the birds were adapted to the keeping conditions.

Fourteen hens of the experimental (AI) group were inseminated with fresh diluted semen 3 times a week, while as a control, 14 hens and 2 cockerels were mated naturally without disturbances. Sperm donor cockerels were kept in the same building but in individual cages under similar condition to the females. Sperm collections were carried out directly before inseminations by the massage method (Burrows and Quinn, 1937). The pooled semen samples were diluted with PBS solution to a concentration of 300 million spermatozoa/female. High doses of semen were used for certain fertilization of all eggs. During 8 weeks, eggs were collected and weighed then incubated for 5 days. After determination of embryonic status (live, dead, developmental stages) yolk samples from embryo-deprived eggs were frozen for analyzing the corticosterone content by RIA.

METHOD OF YOLK CORTICOSTERONE ANALYSIS

The thawed, homogenized yolk samples were extracted (20 min) with dichlor-methane three times. To reduce the quantity of lipids 50-150 µl, 1% Triton-X-100 was added to yolk samples before their extractions (Szőke *et al*, 2005). A 2.5 mg aliquot of yolk samples was used in triplicate for the determination of steroids. Determination of corticosterone (B) equivalents by RIA was carried out according to the method by K. Mihály (Szőke *et al*, 2005). Antibody was developed against Corticosterone -3-CMO-BSA. The sensitivity of the B assay was 10 pg/tube, the intra- and interassay reproducibility were 6–9 and 11–14% in CV. The bound radioactivity (cpm) was measured by LKB-WALLACK scintillation spectrometer. To evaluate the steroid recovery of the extractions and lipid emulsions 20µl (4000 cpm/g yolk) 3H-B was added to the samples and 1/5 part of faeces extracts were used to calculate recovery, which was 73.41±4.85 %. Data were compared using ANOVA F-test and paired t-tests ($p \leq 0.05$). All statistical analysis was done with NCSS 6.0 for Windows.

DETERMINATION OF STATUS OF THE EMBRYO

After 5 days incubation, eggs were opened and visually checked for fertility and the status of embryos in fertile eggs. The germinal discs of visually infertile eggs were dissected out and stained with propidium iodide in order to demonstrate the rate of true infertility (Liptói *et al.*, 2004). Dead embryos and embryos with abnormal developments were phenotypically classified according to Abbot and Yee (1975), as PD (positive development) BWE (blastoderm without embryo), D1-4 (day of embryonic death). The status of live embryos was tested according to their developments, in sizes, lengths and positions.

Results and discussion

CHANGES IN EMBRYONIC DEVELOPMENT

In the stressed (AI) group the ratios of live, normal embryos and dead or abnormal embryos are presented in (Fig 1.). In the first week a low rate of dead or abnormal embryos (1.78%) was observable. It is known that largest quantity of yolk is incorporated into the follicle during the last 72 hours before ovulation (Romanoff and Romanoff, 1949), therefore the effect of corticosterone transfer into the yolk could not have occurred. On the 2nd to the 5th weeks this rate increased to 15-23 % then it decreased to the normal level on the 6th to 7th weeks, when the embryos in all eggs were live and showed normal development. The values of the control group are represented in Table 1. In this group the rate of dead or abnormal embryos fluctuated between 0-2 % during the whole experimental period.

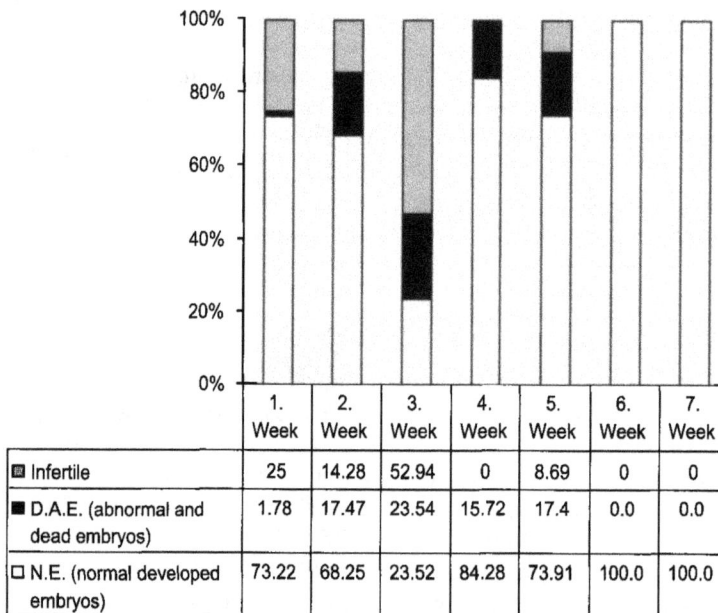

	1. Week	2. Week	3. Week	4. Week	5. Week	6. Week	7. Week
▨ Infertile	25	14.28	52.94	0	8.69	0	0
■ D.A.E. (abnormal and dead embryos)	1.78	17.47	23.54	15.72	17.4	0.0	0.0
☐ N.E. (normal developed embryos)	73.22	68.25	23.52	84.28	73.91	100.0	100.0

Fig 1. Embryonic status after 5 days of incubation in the AI group

Table 1 Percentages of fertile (containing normal, abnormal or dead embryos) and infertile eggs in the control group

Weeks	N.E.(embryos showing normal development) %	D.A.E.(dead and abnormal embryos) %	Infertile eggs %
1	97	1.75	1.25
2	97.08	1.46	1.46
3	98.75	0	1.25
4	97.52	1.65	0.83
5	98.51	1.49	0
6	98.68	0	1.32
7	98.55	1.45	0

The result supports Hayward and Wingfield's (2004) demonstrating that the chicks of corticosterone-implanted mothers hatched worst. In guinea fowl Biczó *et al*, (2004) also found that the effect of handling stress resulted in lower hatching rate. Similarly, in mallards, the handling stress also caused an increase of early embryonic mortality (Szőke *et al* 2003, 2004b.)

CHANGES IN EGG SIZE

One effect of artificial insemination was that the average weight of eggs increased by 1 gram (Table 2), however, the eggs containing abnormal embryos increased by 4-5 grams. The larger egg size manifested in the increase of diameter of the eggs and, additionally, in the increase of the yolk mass (Table 3). The result supports the conclusion of Etches (1995) that the increase of the egg weight is in proportion to increase of the yolk. In the eggs containing normal embryo there were no significant differences in egg weight in the groups. (Table 2, 3).

Table 2. Average weight of eggs in the first 7 weeks in the AI and control group

	Control group	AI group
Egg weight (g)	61.72±2.71(n=100)	62.99±5.06 (n=301)

Table 3. Average weight of eggs containing normal and abnormal embryos in the first 7 weeks in the AI group

	Eggs containing normal embryos in AI group of NE (AI)	Eggs containing normal and dead embryos in AI group of DAE (AI)
Egg weight (g)	60.33±4.69 (n=164)	65.51±5.19 (n=38)

In previous studies with mallard (Szõke *et al*, 2004 a) and guinea fowl (Biczó *et al*, 2004) it was also proved that the handling stress is always accompanied by the increase of the egg mass. In addition, these data support the earlier observation of Pietz *et al*, (1993) on wild mallard as well.

CHANGES IN CORTICOSTERONE (B) CONTENT OF EGG YOLK

Parallel with the egg-weight, artificial insemination significantly ($p< 0.01$) increased the glucocorticoid level of the yolks. Moreover, in eggs with abnormal or dead embryos, the corticosterone level was even higher ($p=0.05$) than in eggs with a normal, live embryo (Fig2) In the control group the correlation between egg weight and yolk corticosterone concentration was 0.55; in the NE (AI) group it was 0.64 and finally the highest correlation (0.74) was in the group DAE (AI) (Fig3-5). In consideration of total egg number, a positive correlation (0.66) was found between the egg weight and the increased glucocorticoid level of the yolks (Fig 6). The higher level of embryo mortality found in the large size egg and was probably the consequence of the higher corticosterone concentration of those eggs, since we could justify relationship between dead or abnormal embryos and high corticosterone concentration as well. Consequently, a relationship was also found between the large size eggs and dead or abnormal embryos. In a previous study on mallards, the effect of handling stress increased the faecal B concentration of mothers and subsequently increased quantity of B deposited into the yolk in addition to increased egg weights (Szõke *et al*, 2004). Similar results were found in side-blotched lizards (Sinervo and DeNardo, 1996), while in Zebra finches the chronically elevated plasma corticosterone had no effect on egg size (Salvante and Williams, 2003).This finding suggests that corticosterone may play different roles in the mediating of egg size in different species.

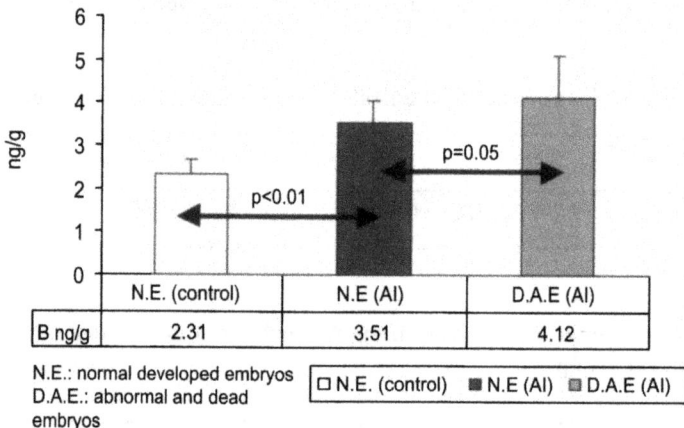

B ng/g	N.E. (control)	N.E (AI)	D.A.E (AI)
	2.31	3.51	4.12

N.E.: normal developed embryos
D.A.E.: abnormal and dead embryos

☐ N.E. (control) ■ N.E (AI) ▨ D.A.E (AI)

Fig 2. Yolk corticosterone concentrations in the control and AI groups

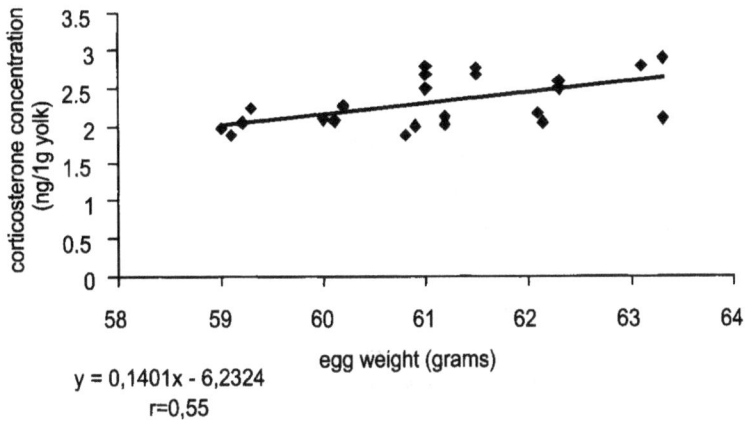

$y = 0,1401x - 6,2324$
$r=0,55$

Fig 3. Correlation between the egg weight and corticosterone level in the yolk of control group (NE(control))

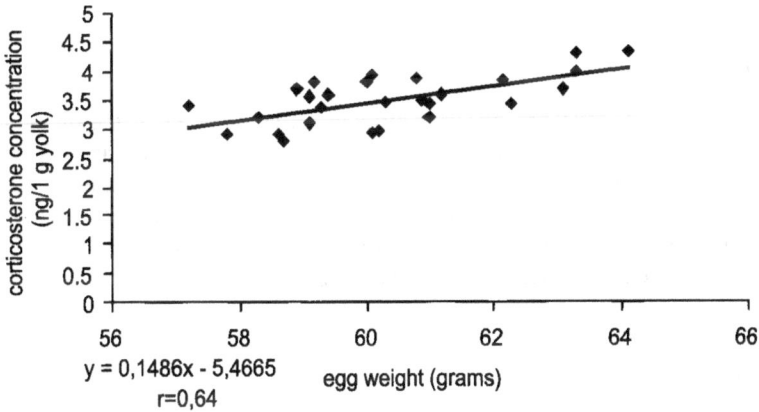

$y = 0,1486x - 5,4665$
$r=0,64$

Fig 4. Correlation between the egg weight and corticosterone level in eggs containing yolk of normal embryos of AI group (NE(AI))

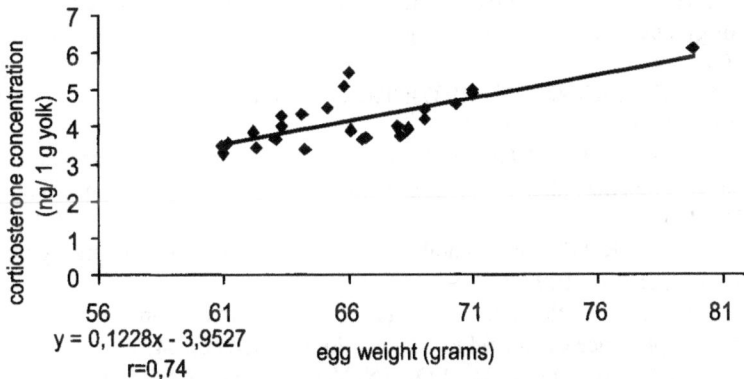

$y = 0,1228x - 3,9527$
$r=0,74$

Fig 5. Correlation between the egg weight and corticosterone level in eggs containing yolk of abnormal and dead embryos of AI group (DAE (AI))

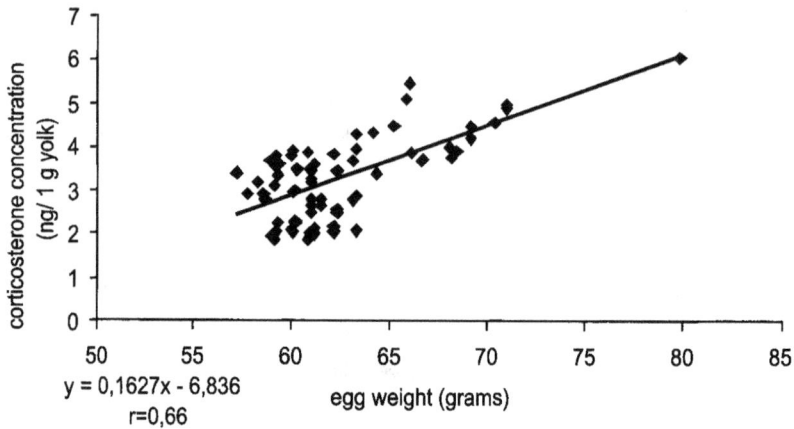

y = 0,1627x - 6,836
r=0,66

Fig 6. Correlation between the egg weight and yolk corticosterone level of total number of eggs

Conclusions

The study confirmed earlier results from different bird species where the stress on the mother – in this case the practice of artificial insemination – can influence the yolk corticosterone content and subsequently the egg weight and the embryonic development in domestic fowl as well. Regarding the adaptation ability of hens, the alteration of these parameters turned back to the regular (basic, physiological) level after a while. According to the study, 6 weeks are needed for the adaptation of hens to the AI procedure.

References

ABBOT, U.K., YEE, G.W. (1975): Avian Genetics. In *Handbook of Genetics*. V.4.R.C. King, Ed. Plenum Publ. Corp. NY. 151-200.

ARCOS, M. (1972): Steroids in egg yolk. Steroids 19:, 25-34.

BICZÓ, A., SZŐKE, ZS. and PÉCZELY, P. (2003): Plasma and fecal sexualsteroid parameters of guinea fowl (Numida meleagris). Hungarian Journal of Animal Production 52: 189-194

BURROWS, W.H., QUINN, J.P. (1937): The collection of spermatozoa of domestic fowl and turkey. Poult. Sci. 16:19-24.

HAYWARD, L.S. AND WINGFIELD, J.C. (2004): Maternal corticosterone is transferred to avian yolk and may alter offspring growth and adult phenotype. Gen. Comp. Endorinol. 135(3): 365-71.

HERRENKOHL, L.R. (1979): Prenatal stress reduces fertility and fecundity in female offspring. Science 206: 1097-1099.

HERTELENDY, F. & COMMON, R. H. (1965): A chromatographic investigation of egg yolk for the presence of steroid estrogens. Poultry Science 44, 1205-1209.

LIPAR, J. L., KETTERSON, E. D., NOLAN, V. & CASTO, J. M. (1999): Egg yolk layers vary in the concentration of steroid hormones in two avian species. Gen. Comp. Endocrinol., 115: 220-227

LIPTÓI, K., VARGA, A., HIDAS, A., BARNA, J. (2004): Determination of the rate of true fertility in duck breeds by the combination of two *in vitro* methods. Acta Vet. Hung. 52 (2): 227-233.

MCCORMICK MI (1999): Experimental test of the effect of maternal hormones on larval quality of coral reef fish. Oecologia. 118: 412-422

MCNABB, F. M. A. & WILSON, C. M. (1997): Thyroid hormone deposition in avian eggs and effects on embryonic development. American Zoologist, 37:553-560.

MORGAN MJ, WILSON, CE, CRIM, LW (1999): The effect of stress on reproduction in Atlantic cod. J. Fish Biol. 54: 477-488.

MOUSSEAU, T.A. FOX, C.W. (1998): Maternal effects as adaptation. Oxford University Press. New York

PIETZ, P.J., KRAPU, G.L., GREENWOOD, R.J. and LOKEMOEN, J.T. (1993): Effects of harness transmitters on behavior and reproduction on wild Mallards. Journal of Wildlife Management 57(4): 696-703

POTTINGER, TG, CARRICK TR (2000): Indicators of reproductive performance in rainbow trout *Oncorhynchus mykiss* (Walbaum) selected for high and low responsiveness to stress. Aquac. Res. 31: 367-375.

RIDDLE, O. & DUNHAM, H.H. (1942): Transformation of males to intersexes by estrogen passed from blood of ring doves to their ovarian eggs. Endocrinol. 30: 959-968.

ROMANOFF, A. L. & ROMANOFF, A. J. (1949): The Avian Egg. Academic Press. New York.

SALVANTE, KG, WILLIAMS, TD (2003): Effects of corticosterone on the proportion of breeding females, reproductive output and yolk precursor levels. Gen. Comp. Endocr. 130:205-214.

SCHWABL, H. (1993): Yolk is source of maternal testosterone for developing birds. Proc. Natl. Acad. Sci U.S.A, 90: 11446-11450.

SINERVO AND DENARDO (1996): Costs of reproduction in the wild: path analysis of natural selection and experimental tests of causation. Evolution 50: 1299-1313.

SZŐKE, ZS, FERENCZI, SZ, ÁDÁM, D, BICZÓ A, PÉCZELY P (2003): Effect of maternal stress on yolk steroids and on somatic parameters of offspring in mallard (*Anas plathyrynchos*) Hung. Journal of Animal Production 52.(2): 180-188.

SZŐKE, ZS , BICZÓ, A , PÉCZELY, P (2004a): Effect of handling stress on the egg production and some egg-parameters Proc. 8th International Symposium on Avian Endocrinology, Phoenix AZ, USA, P77

SZŐKE, ZS , FERENCZI, SZ, BICZÓ, A, PÉCZELY, P (2004b): Effect of maternal handling stress on the steroid deposition into the yolk and on the offspring Proc. 8th International Symposium on Avian Endocrinology, Phoenix AZ, USA, P78

SZŐKE, ZS., BICZÓ, A., BARNA, J. PÉCZELY, P. (2005): Determination of steroid hormones of egg yolk as a diagnostic tool for determination of the effects of stress and maternal investments. (In Hungarian) Proc. of Conference on "Reproduction biology in wild animal, breeding programs in zoos" Budapest, Hungary, pp.18-20.

TAKAHASHI, L.K., HAGLIN. C., KALIN, N.H. (1992): Prenatal stress potentiates stress-induced behavior and reduces the propensity to play in juvenile rats. Physiol. Behav. 51: 319-323.

WARD I.L. (1972): Prenatal stress feminizes and demasculinizes the behavior of males. Science. 175: 82-84.

Using un-traditional sources of antioxidants to cope with stress in broiler chickens

EID, Y. Z. [1]; EBEID, T. A. [1]; OHTSUKA, A [2] AND HAYASHI, K [2]

[1]Department of Poultry Production, Kafr El-sheikh Faculty of Agriculture, Tanta University, 33516 Kafr El-sheikh, EGYPT; [2]Department of Biochemical Science and Biotechnology, Faculty of Agriculture, Kagoshima University, 1-21-24 Korimoto, Kagoshima 890-0065, JAPAN

The present study was conducted to compare the antioxidant activities of un-traditional antioxidants; epigallocatechin gallate (Epi-Galo) extracted from green tea, crude extract of tea polyphenols (PP), and vitamin E (Vit. E), under the oxidative stress induced by corticosterone (CTC) administration in broiler chickens. Chicks (Cobb strain) at 12 days of age were divided into four groups [control, Epi-Galo (20 mg/ kg diet), PP (1.0%/ kg diet) and Vit. E (300 mg α-tocopheryl acetate/ kg diet)]. Each group consisted of non-CTC (-) group and CTC (20 mg/kg diet) (+) group, the birds were given the experimental diets from 16 to 27 days of age _ad libitum._ (n = 12). At the end of the experimental period CTC treatment significantly reduced body weight gain and PP also reduced body weight gain while other antioxidants had no effect. Abdominal fat deposition and plasma cholesterol were significantly increased by CTC and this increment was suppressed by the antioxidants, especially by PP. However, Epi-Gallo and PP reduced plasma cholesterol in the birds of CTC (-) group. Liver TBARS was increased by CTC and significantly decreased by the antioxidants, especially by PP. The activity of liver glutathione peroxidase (GSH-Px) was reduced by CTC and it was minimized by the antioxidants with a significant effect only by PP. The effect on sodium superoxide dismutase (SOD) was similar. These results clearly demonstrate the effectiveness of the un-traditional antioxidants used to minimize the stress in broiler chickens. However, the levels of the antioxidants are needed to be evaluated in the future.

Keywords: antioxidants; catechins; oxidative stress; stress markers; broiler

Introduction

Welfare is one of the issues which affect the quantity and quality of poultry products, which in turn affect human health. Stress (transportation, electric shocks, fasting, low or high ambient temperature, etc.) is one of the limiting factors which affect the welfare and the production of poultry. Stress induces free radical formation and develops oxidative stress, initiating lipid peroxidation in tissues followed by abnormalities in lipid metabolism, growth and production in general.

*Corresponding author: Dr. Yahya Zakaria Eid
E-mail: yahyaze@gmail.com

Under stressful conditions, antioxidant requirement is thought to be increased to protect polyunsaturated fatty acids from peroxidation. It works to scavenge free radicals and acts as a terminator of lipid peroxidation (Taniguchi *et al.*, 1999, Eid *et al.*, 2003). Administration of vitamin E under oxidative stress induced by CTC resulted in reduction of free radicals formation followed by decreases in lipid peroxidation and cholesterol level in plasma (Bjorneboe *et al.*, 1994).

Polyphenols are found in many plants such as fruits, vegetables and tea etc. Tea (*Camellia sinensis*) is the most widely consumed beverage in the world and its polyphenols are known as catechins. Catechins are water-soluble polyhydroxylated flavonoids. The catechins that are observed in tea are (-)-epicatechin (EC), (-)-epigallocatechin (EGC), (-)-epicatechin gallate (ECG), and (-)- epigallocatechin gallate (EGCG), which differ in the number and position of the hydroxyl groups in the molecule (Douglas *et al.*, 1997; Ninomiya *et al.*, 1997; Matthew and Douglas 1997). Recently, polyphenols have been used as un-traditional source of antioxidants. A connection has been suggested between the consumption of polyphenol-rich foods or beverages and protection against cancer, coronary heart disease, inflammation and reducing lipid oxidation (Morre *et al.*, 2000; Veron *et al.*, 2000).

The present study was designed to study the stress response of dietary epigallocatechin gallate, polyphenol and Vit. E under oxidative stress conditions induced by CTC administration in broiler chickens.

Material and methods

One-day-old male broiler chickens (Cobb strain) were provided with water and a basal diet (20% crude protein and metabolizable energy ME 3,250 Kcal/Kg) *ad libitum* for the first 12 days. The experiments were conducted in a controlled room with 14h light: 10h dark cycle, 25 ° C temperatures and the relative humidity was 50-70% throughout the experiment.

Experimental diets and feeding: On day 12, 96 birds were housed in individual cages and divided into four groups {control, Epi-Galo (20 mg Epigallocatechin gallate/ kg diet), PP (1.0% Polyphenol/ kg diet and Vit. E (300 mg α-tocopheryl acetate/ kg diet)} and each group was further divided into subgroups, 0 (-) and 20 (+) mg CTC /kg diet (n= 12). The birds were given the experimental diets from 16 to 27 day of age *ad libitum*.

Chemical analysis: At the end of the experimental period (day 27), all birds were killed by decapitation. Blood and hepatic samples were collected for chemical analysis. Lipid peroxidation in the liver was assessed as the concentration of thiobarbituric acid reactive substance TBARS (Richard *et al.*, 1992). Cholesterol was measured by an enzymatic assay using a commercial kit (Wako Pure Chemical Industries, Ltd., Osaka, Japan). The hepatic glutathione peroxidase (GSH-Px) activity was determined by the method of Levander *et al.* (1983). The superoxide dismutase (SOD) activity was measured according to Mirro *et al.* (1989).

Statistical analysis: Data were analyzed by two-way analysis of variance (ANOVA) using the General Liner Model procedure of the statistical analysis system software package (SAS Institute Inc., Cary, NC, USA, 1988) with Duncan's multiple-range test. P-value ≤ 0.05 was considered to be statistically significant.

Results and discussion

CTC treatment significantly reduced body weight gain and PP also reduced body weight gain while other antioxidants had no effects (Figure 1A). Many studies showed the positive effect of

CTC in growth inhibition, this effect could be explained by decreased muscle protein synthesis and enhanced muscle proteolysis (Hayashi *et al.*, 1994). Skeletal muscle proteins may be damaged by oxidative stress as reported by (Hunt *et al.*, 1988). Moreover, Eid *et al.* (2003) showed that high level of PP could enhance the growth inhibition resulted from the oxidative stress induced by CTC in broiler chickens. However, PP in normal condition may cause growth inhibition. This result is in consistent with the results reported by Biswas and Wakita (2001).

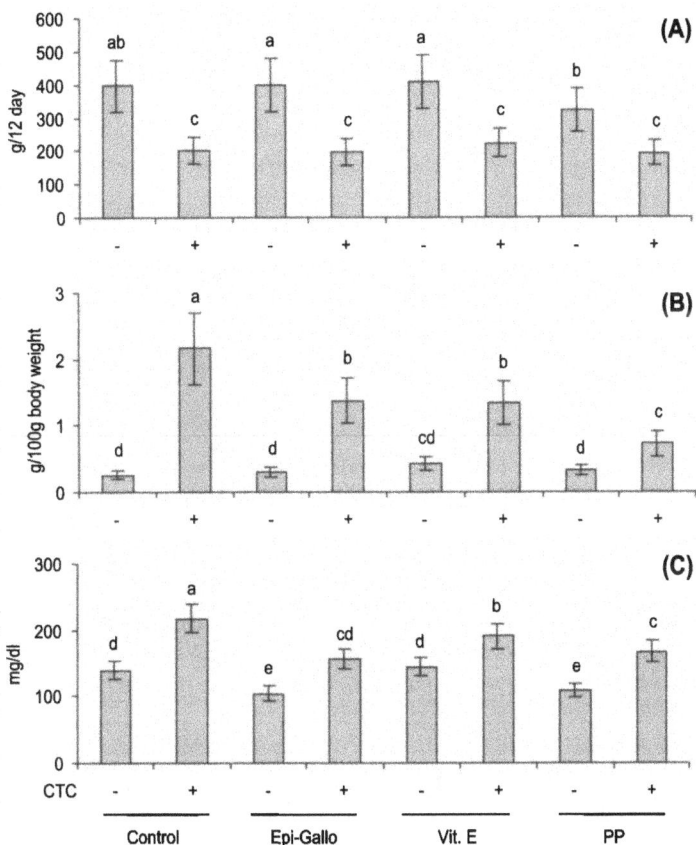

Figure 1. Effect of oxidative stress and antioxidants on (A) body weight gain, (B) abdominal fat deposition and (C) plasma cholesterol level. Values are expressed as means ± standard deviation (STD); means with different script are significantly different from each other (P = 0.05).

The results presented in Figure 1 (B, C) indicated that abdominal fat deposition significantly increased by CTC and this increment was suppressed by the antioxidants, especially by PP, and the effects on plasma cholesterol were similar. However, Epi-Gallo and PP also reduced plasma cholesterol in the birds of CTC (-) groups. It is known that glucocorticoids cause hyperglycemia, hypertriglyceridemia and hypercholesterolemia, which are companied by increases in lipid peroxidation (Hidalgo *et al.*, 1988). These results suggesting that the antioxidants play an important role in lipid metabolism of the whole body (Taniguchi *et al.*, 2001; Eid *et al.*, 2003). This strong effect of tea catechin could be explained by its inhibitory effect on intestinal lipid absorption followed by a decrease of abdominal fat content. Tea catechin may also change the formation of micell that mediates re-absorption of bile acid (Biswas and Wakita 2001).

As shown in Figure (2A), liver TBARS was increased by CTC and PP significantly decreased by the antioxidants, especially by PP. The activity of liver glutathione peroxidase (GSH-Px) was reduced by CTC and this minimized by the antioxidants with significant effect only by PP (*Figure, 2B*). The effect on sodium superoxide dismutase (SOD) was similar (*Figure, 2C*). These findings are in agreement with our previous work (Eid *et al.*, 2003). It is also known that epigallocatechin gallate strongly inhibits cholesterol ester and apolipoprotein B-100 fragmentation of LDL (Moure *et al.*, 2001). Many authors reported the positive effect of green tea or green tea catechin in enhancing the activity of these enzymes in rats or in human hepatic cell culture (Murakami *et al.*, 2002). Polyphenolic compounds proved to be more effective than Vit. E as a free radical scavenging (Sanchez-Moreno *et al.*, 1999). Furthermore, green tea catechin proved to inhibit the reduction of Vit. E in chicken meat. That was positively dose dependant with the level of catechin (Tang *et al.*, 2002). This effect of Vit. E protection is due to dietary catechins may partly contribute to the strong antioxidant activity of tea catechin *in vivo*.

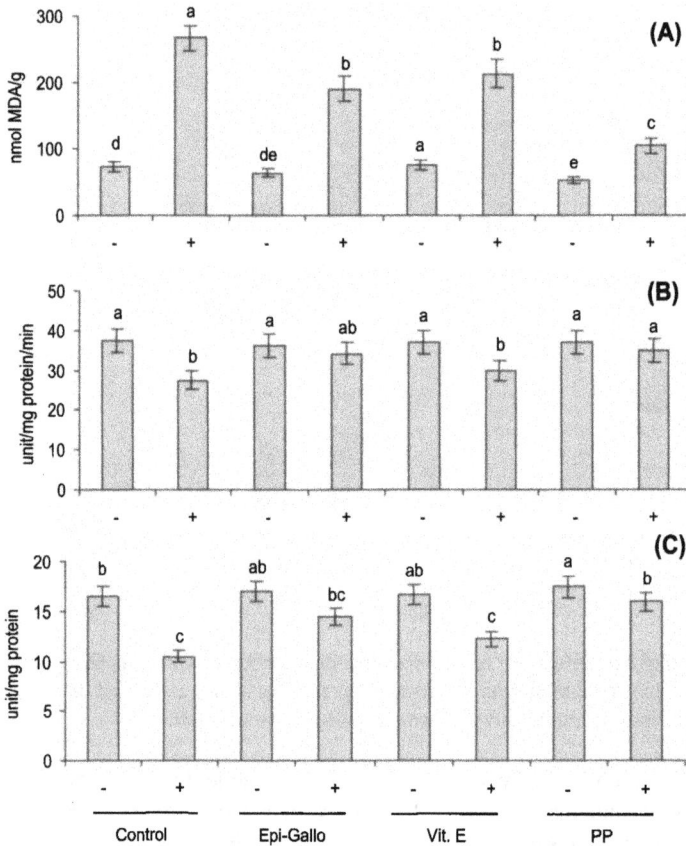

Figure 2. Effect of oxidative stress and antioxidants on (A) hepatic TBARS, (B) hepatic GSH-Px and (C) hepatic SOD levels. Values are expressed as means ± standard deviation (STD); means with different script are significantly different from each other (P = 0.05).

Based on the pervious results, it could be mentioned that the antioxidants play an important role under the oxidative stress condition by decreasing lipid peroxidation level. This could be translating to a normal and better performance, according to the low effect of free radicals in the presence of the antioxidants. Tea polyphenols are effective in reducing abdominal fat deposition, plasma cholesterol, and enhancing the activity of the hepatic antioxidative enzymes under oxidative stress condition. These natural and un-traditional antioxidants could be contributed in enhancing the welfare of the raised birds under stressful conditions. However, graded levels of tested antioxidant are needed to be investigated.

References

BISWAS, MD. AH. and WAKITA, M. (2001) Effect of dietary Japanese green tea powder supplementation on food utilization and carcass profiles in broilers. Journal of Poultry Science **38**: 50-57.

BJORNEBOE, A., BJORNEBOE, G-E. AA. and DERVON, C. A. (1994) Absorption, transport and distribution of vitamin E. Journal of Nutrition **120**: 233-242.

DOUGLAS, A. B., SHEILA, A. W. and LIESBETH, M. B. 1997. The chemistry of tea flavonoids. Critical Review Food Science and Nutrition **37**: 693-704.

EID, Y., OHTSUKA, A. and HAYASHI, K. (2003) Tea polyphenols reduce glucocorticoid induced growth inhibition and oxidative stress in broiler chickens. British Poultry Science **44**: 127-32.

HAYASHI, K., NAGAI, Y., OHTSUKA, A. and TOMITA, Y. (1994) Effects of dietary corticosterone and trilostane on growth and skeletal muscle protein turnover in broiler cockerels British Poultry Science **35**:789-798.

HIDALGO, J., CAMPMANY, L., BORRAS M, GARVEY, J. S. and ARMARIO, A. (1988) Metallothionein response to stress in rats: role in free radical scavenging. American Journal of Physiology **255**: E518-524.

HUNT, J., DEAN, R.T. and WOLFF, S. P. (1988) Hydroxyl radical production and autoxidative glycosylation. Biochemistry Journal **256**: 205-212.

LEVANDER, O. A., DELOACH, D. P., MORRIS, V. C., and MOSER, P. B. (1983) Platelet glutathione peroxidase activity as an index of selenium status in rats. Journal of Nutrition **113**: 55-63.

MATTHEW, E. H. and DOUGLAS, A. B. (1997) Tea chemistry. Critical Review in Plant Science **16**: 415-480.

MIRRO, R., ARMSTEAD, W. M., MIRRO, J. J., BUSIJA, D.W., and LEFFLER, C. W. (1989) Blood induced superoxide anion generation on the cerebral cortex of newborn pigs. American Journal of Physiology **257**: H1560-H1546.

MORRE, D. J. BRIDGE, A., WU L. and MOREE, D. M. (2000) Preferential inhibition by (-)epigallocatechin-3-gallate of the cell surface NADH oxidase and growth of transformed cells in culture. Biochemical Pharmacology **60**: 937-946.

MOURE, A., CRUZ, J. M., FRANCO, D., DOMINGUEZ, J. M., SINERIRO, J., DOMINGUEZ, H., NUNEZ, M. J. and PARAJO, J. C. (2001) Natural antioxidants from residual sources. Food Chemistry **72**: 145-171.

MURAKAMI, C., HIRAKAWA, Y. INUI, H., NAKANO, Y. and YOSHIDA, H. (2002) Effects of Epigallocatechin 3-O-gallate on cellular antioxidative system in HepG2 cells. Journal of Nutritional Science and Vitaminology **48**: 89-94.

NINOMIYA, M., UNTEN, L. and KIM, M. (1997) Chemical and physicochemical properties of green tea polyphenols, In: Yamamoto, T.; Juneja, L. R. & Kim, M. (Eds) *Chemistry and applications of green tea*, pp. 23-36 (CRC press LLC).

RICHARD, M. J., PORTAL, B., MEO, J., COUDRAY, C., HADJIAN, A. and FAVIER, A. (1992) Malondialdehyde kit evaluated for determining plasma and lipoprotein fractions that react with thiobarbituric acid. Clinical Chemistry **38**: 704-709.

SANCHEZ-MORENO, C., AND LARRAURI, J. A. and SAURA-CALIXTO, F. (1999) Free radicals scavenging capacity and inhibition of lipid oxidation of wines, grape juices and related polyphenolic constituents. International Food Research **32**: 407-412.

SAS institute INC. (1988) SAS User's guide: Statistics. SAS Institute, Cray, NC, USA.

TANG, S. Z., KERRY, J. P., SHEEHAN, D., BUCKLEY, D. J. and MORRISSEY, P. A. (2002) Dietary tea catechins and iron-induced lipid oxidation in chicken meat, liver and heart. Meat Science **56**: 285-290.

TANIGUCHI, N., OHTSUKA, A. and HAYASHI, K. (2001) A high dose of vitamin E inhibits adrenal corticosterone synthesis in chickens treated with ACTH. Journal of Nutritional Science and Vitaminology **47**: 40-46.

TANIGUCHI, N., OHTSUKA, A. and HAYASHI, K. (1999) Effect of dietary corticosterone and vitamin E on growth and oxidative stress in broiler chickens. Animal Science Journal **70**: 195-200.

VERON, E. S., GRAY, J. K., DOUGLAS, B., CHARLES, W. B., RAJENDRA, M., DONYA, B., CAROLINE, C. S., SONGYUN, Z. and SHEELA, S. (2000) Comparative chemopreventive mechanisms of green tea, black tea and selected polyphenol extracts measured by *in vitro* bioassays. Carcinogenesis **21**: 63-67.

www.ingramcontent.com/pod-product-compliance
Lightning Source LLC
Chambersburg PA
CBHW061813210326
41599CB00034B/6985